After all—what could you possibly learn from a cheerleader?

GO GEEK. NOT GREEK.

Choose individuality not conformity. Imagination over social acceptance. Learning, not just going to class. Technology, not technician. UAT offers degrees in Game Design, Digital Animation, Artificial Life Programming, Digital Video, Web Design, Computer Forensics, Network Security, Software Engineering, Technology Management, Game Programming and more.

www.uat.edu > 800.658.5744

Make:
technology on your time

Volume 05

TRASHCAN

40

VOYAGE TO THE BOTTOM OF THE POOL Students from universities and trade schools in the U.S., Canada, and India entered the 8th International Autonomous Underwater Vehicle Competition. At stake was $20,000 in prize money, which considering the cost and time to create a submarine, isn't the motivating factor. It's all about the bragging rights.

INTEL'S IN-HOUSE MAKER Eric Paulos is a computer scientist at Intel Research Berkeley, a corporate "lablet" in the penthouse of a down-town office building where a dozen or so scientists explore the edges of computing and networking technology. For Paulos, every city is an ideal laboratory to test what he calls "objects of wonderment."

Vol. 05, February 2006. MAKE (ISSN 1556-2336) is published quarterly by O'Reilly
Media, Inc. in the months of February, May, August, and November. O'Reilly Media
is located at 1005 Gravenstein Hwy North, Sebastopol, CA 95472, (707) 827-
7000. SUBSCRIPTIONS: Send all subscription requests to MAKE, P.O. Box 17046,
North Hollywood, CA 91615-9588 or subscribe online at makezine.com/offer or
via phone at 866-289-8847 (U.S. and Canada), all other countries call (818) 487-
2037. Subscriptions are available for $34.95 for 1 year (4 quarterly issues) in the
U.S. Canada: $39.95 USD; all other countries: $49.95 USD. Application to Mail at
Periodicals Postage Rates is Pending at Sebastopol, CA and at additional mailing office
POSTMASTER: Send address changes to MAKE, P.O. Box 17046, North Hollywood, CA
91615-9588.

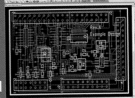
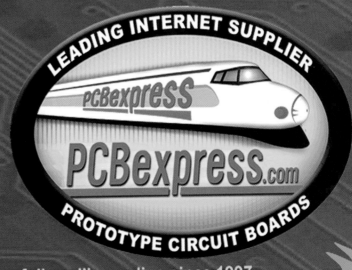

Make: Projects

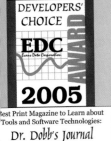

Volume 05

154

ON THE COVER
Most people think of water
rockets as being those cheap
red and white plastic toys. But
a thriving community of water
rocket enthusiasts have
launched their hand-built
creations to heights of nearly
1,000 feet. Our project, which
starts on page 78, comes with
a parachute for a soft landing.
Photograph by Topher Lucas

DIY 115

Make:
technology on your time™

EDITOR AND PUBLISHER
Dale Dougherty
dale@oreilly.com

EDITOR-IN-CHIEF
Mark Frauenfelder
markf@oreilly.com

CREATIVE DIRECTOR
David Albertson
david@albertsondesign.com

MANAGING EDITOR
Shawn Connally
shawn@oreilly.com

ART DIRECTOR
Kirk von Rohr

ASSOCIATE EDITOR
Phillip Torrone
pt@makezine.com

DESIGNERS
Sarah Hart
Kristy McKoy

PROJECTS EDITOR
Paul Spinrad
pspinrad@makezine.com

SPECIAL PROJECTS EDITOR
David Pescovitz

ASSOCIATE PUBLISHER
Dan Woods
dan@oreilly.com

ASSISTANT EDITOR
Arwen O'Reilly

CIRCULATION DIRECTOR
Heather Harmon

COPY CHIEF
Goli Mohammadi

ADVERTISING COORDINATOR
Jessica Boyd

MARKETING & EVENTS COORDINATOR
Rob Bullington

COPY EDITORS/RESEARCH
Terry Bronson
Keith Hammond

ONLINE MANAGER
Terrie Miller

MAKE TECHNICAL ADVISORY BOARD:
**Gareth Branwyn, Joe Grand,
Saul Griffith, William Gurstelle, Bunnie Huang,
Tom Igoe, Mister Jalopy, Steve Lodefink**

PUBLISHED BY O'REILLY MEDIA, INC.
Tim O'Reilly, CEO
Laura Baldwin, COO

Visit us online at makezine.com
Comments may be sent to editor@makezine.com

For advertising and sponsorship inquiries, contact:
Dan Woods, 707-827-7068, dan@oreilly.com

Contributing Artists:
Howard Cao, Roy Doty, Paul Hansen, Chad Holder, Dustin Amory Hostetler, Christopher Hujanen, Tim Lillis, Topher Lucas, Jenny Pfeiffer, N55, Emily Nathan, Eric Rife, Nik Schulz, Anne Sigler, Damien Scogin, David Sobo

Contributing Writers:
Tim Anderson, Thomas Arey, Dave Battino, Joost Bonsen, Gareth Branwyn, Abe Connally, Cory Doctorow, Nick Dragotta, George Dyson, Jonathan Foote, Joe Grand, Saul Griffith, William Gurstelle, Keith Hammond, Larry Harmon, Rob Hartman, Christopher Holt, Bunnie Huang, Tom Igoe, Andy Ihnatko, Mister Jalopy, Stefan Jones, Chris Kentworthy, Jason Kohrs, Mark Lengowski, William Lidwell, Steve Lodefink, Robert Luhn, Merlin Mann, Dave Mathews, Josie Moores, Danny O'Brien, Tim O'Reilly, Ross Orr, Tom Owad, Bob Parks, David Pescovitz, Jenna Phillips, Dan Picard, Charles Platt, Julie Polito, Michael Pryor, Douglas Repetto, Matthew Russell, Brian Sawyer, Bob Scott, H.B. Siegel, Ewan Spence, Bruce Sterling, Bruce Stewart, Cy Tymony, Steve Vigneau, Howard Wen

Interns: Adrienne Foreman (web), Jake McKenzie (eng.), Ty Nowotny (eng.), Aly Rasmussen (design)

MAKE is printed on recycled paper with 10% post-consumer waste and is acid-free. Subscriber copies of MAKE Volume 05 were shipped in recyclable plastic bags.

Contributors

Tim Lillis (*Wind Powered Generator* illustrations) is still waiting for the hand-drawn blueprints that he sent to Hasbro as a young lad to be made into Transformers, but in the meantime, he works as a freelance designer and illustrator. A recent arrival to San Francisco, Lillis once worked with Kaiju Big Battel, the world's only Live Monster Fighting Tournament, making monster costumes, miniature cityscapes, and giant cartoonish props out of foam, and acting as a producer for their live events. Tim's hobbies include ruining karaoke, making way too many puns, and designing whatever he can at narwhalcreative.com. He loves AMC Eagles, the best car ever.

Cy Tymony (*Sneaky Uses*) reads six newspapers a day. "I am inquisitive about all aspects of how life works," the Los Angeles author/inventor says. His career started as a child, when he made a sleeve-mounted shocker device to defend himself from schoolyard bullies. His *Sneaky Uses* books explore the delight of finding. He also roller-skates, practices aikido and judo, plays with cardboard boomerangs, and eats tuna salad when he is not working on finding still more sneaky uses for everyday things.

Jenny Elia Pfeiffer (*Woody's World* photos) is a San Francisco-based artist and photographer. Her portraits and travel photography have appeared in several local and national publications. She also collaborates with her sister, Lisa Pfeiffer, on multimedia works that have been featured in galleries in San Francisco and Prague. Her favorite food is ice cream (Oreo cookie).

George Dyson (*Treehouse*) is a historian of technology whose interests range from the development (and redevelopment) of the Aleut kayak to the evolution of digital computing, telecommunications, and nuclear bomb-propelled space exploration. Dyson, who lives in Bellingham, Wash., divides his time between building boats and writing books. He considers himself a follower of Nathaniel Bishop, who, while paddling a paper kayak to the Gulf of Mexico from Québec in 1874, urged his audience at Princeton University to "seek in his friendly canoe that relief which nature offers to the tired brain. Let him go into the wilderness and live close to his Creator by studying his works."

An inveterate tinkerer and "broad-spectrum hobbyist," **Steve Lodefink** (*Soda Bottle Rocket*) just can't say no to a cool project. At 3, he was already reverse-engineering the peanut butter and jelly sandwich: "I figured out where all of the parts were, found a good tool, and built one. I've been doing it ever since." He lives in Seattle with his wife and two sons, two cats, five tarantulas, and 24 African cichlids, and thinks that one of life's great pleasures is a really sharp aged cheddar cheese. "I'm a simple man," he says. He looks at life's debris at finkbuilt.com.

"Tall and skinny, with an affinity for firecrackers and vinyl records," **Larry Harmon** (*Underwater Vehicles*) has been exploring San Diego's underbelly for years. His fanzine, *Genetic Disorder*, is an ode to anything "that would embarrass our Chamber of Commerce." When not at the library doing research for a book about the history of San Diego's punk rock scene, he plays guitar in the Dissimilars (myspace.com/thedissimilars), "a trashy garage rock band." He spends his days as an editor at a community newspaper, three blocks from the beach.

If **Tom Owad** (*Retrocomputing*) could be any animal, he'd be something from the Galapagos. His interest in extraordinary and incomparable things extends to his human existence as well. He spends his days tinkering and learning, and is the owner and webmaster of applefritter.com, an online Macintosh community of artists and engineers dedicated to the "obscure, unusual, and exceptional." He serves on the board of directors and is webmaster and archivist of the Apple I Owners Club, and is also the author of *Apple I Replica Creation*.

Maker Faire

ON TO YEAR TWO

Dale Dougherty and
Mark Frauenfelder look back
on MAKE's first year.

Q: Now that MAKE is a year old, what have you learned?

Dale Dougherty: Someone once told me that you don't create a magazine — you discover it. You discover an audience, if you're lucky, and then you have to work hard to keep up with them.

Mark Frauenfelder: We hoped there would be an audience for MAKE. The surprising thing is how big that audience is. The world has far more makers than I would have dreamed. It makes me feel good about the human race.

DD: I had an interviewer ask me, "Don't you have to be pretty smart to do the projects in MAKE?" I answered, "I think most people are pretty smart. It's a matter of having the time to make things, and that's true of cooking, photography, electronics, and lots of areas." Often, there's some technical knowledge you need to acquire, but you learn as you go, and you can have lots of fun. It's like playing a game.

MF: The best games are ones that are easy to learn but challenging to master, like chess. I think the same thing can be said for great projects. The kite aerial photography project in our first issue was really simple, giving you a lot of bang for your buck. But more importantly, it pointed the way to a community of photographers who are improving on one another's designs, and building remarkably sophisticated systems.

DD: I guess it's my move now! So one of the big surprises for me in MAKE's first year is hearing from kids or from their parents. I got a note recently from a father who said he was saving back issues of MAKE to do projects with his son when he was old enough. I hope MAKE's projects engage kids of all ages in science and technology.

MF: I agree. There's a reason why we don't label any of the projects as being for kids or for adults. If kids are interested in a complex project, they'll rope in an adult to help them. I'm sure just as many adults will ask their kids to help them with the projects. My 8-year-old daughter is already teaching me how to work the user interfaces on the technology around our house.

DD: When you're a kid and you see something cool, such as the soda-bottle rocket (which is a project in this issue, page 79), you don't just want to watch someone else launch it. You want to do it yourself. You want to know how to build it yourself. That's the maker spirit. It's fun to meet makers and see what they do. It's inspiring. I'm looking forward to our Maker Faire in April.

MF: So am I. The best part of my job is meeting with makers and hearing their stories. One of the things we'll be focusing on more in MAKE this year is the stories behind the projects we feature, and the community of makers involved in them.

DD: Finally, I'd like to thank all those readers who have written us saying how much they love the magazine. I'd also like to thank the team we've put together to make the first year of MAKE such a success. We're looking forward to another exciting year.

Mark Frauenfelder is editor-in-chief and Dale Dougherty is editor and publisher of MAKE.

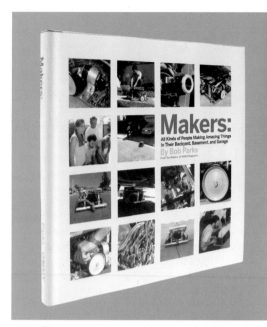

Tim O'Reilly

NEWS FROM THE FUTURE

"THE FUTURE IS HERE. IT'S JUST NOT EVENLY DISTRIBUTED YET"— WILLIAM GIBSON

I HAD LASER EYE SURGERY THE OTHER DAY, and after more than 40 years of wearing glasses so strong that I was legally blind without them, I can see clearly on my own. I had a perfect outcome: 20/20 for far vision, yet still able to read and do other close work as well. I keep saying to myself: I'm seeing with my own eyes!

But in order to remove my need for prosthetic vision, the surgeon ended up relying on prosthetics of her own, performing the surgery with the aid of high-tech equipment and specialized technicians.

First, they mapped my eyes with a device called a corneal topographer, and came up with a modification plan. Then they used a laser to blister the surface of my cornea, and 20 minutes later, the surgeon used a microkeratome to lift the flap of the blister so another laser could do the real mods to the deeper layers of the cornea. During the actual surgery, apart from lifting the flap and smoothing it back into place after the laser was done, her job was to clamp open my eyes, hold my head, utter reassuring words, and tell me, sometimes with urgency, to keep looking at the red light. Afterwards, I asked what would happen if my eyes drifted and I didn't stay focused on the light. "Oh, the laser would stop. It only works when your eyes are tracking."

In short, surgery this sophisticated could never be done by an unaugmented human being. The human touch of my superb doctor was paired with the inhuman accuracy of complex machines, a 21st-century hybrid freeing me from the tyranny of assistive devices first invented in 13th-century Italy.

Whether or not we're heading for a Kurzweil-style singularity, in which humans merge with machines, an increasing number of our activities are only possible with the aid of computers and other complex devices. My eye surgery is only one example.

The revolution in sensors, computers, and control technologies is going to make many of the daily activities of the 20th century seem quaint as, one by one, they are reinvented in the 21st. It's all over the news — remote drone warfare, powered exoskeletons, automatic language translators, cellphones that know their owners by the way they walk — but these high-profile examples tend to obscure the ways that computer technology is remaking the everyday world.

Take the automobile: it's a short step from high-tech vehicles that can only be debugged with the aid of computer diagnostics to vehicles that can only be driven by a computer. "The long run" — the last road rally of those diehards who insist on driving their own vehicles — imagined in the 1989 novel of the same name by Daniel Keys Moran, may not be that far off.

As we appreciate the wonders of technology, let's not forget the dangers of taking them for granted.

For everything that is gained, something is lost. The classic 1958 essay "I, Pencil," by Leonard Read, compellingly makes the point that in a connected world, even simple objects like the humble lead pencil require the expertise and activity of machinery, business processes, and thousands of coordinated people. How true that is today of our complex devices, composed of throwaway components themselves made in multi-billion-dollar factories.

In this world, the role of the maker is especially important, lest we fall into the future foretold nearly a century ago by E.M. Forster in "The Machine Stops," in which people cocooned in their high-tech homes are helpless when their technological assistants go silent.

Knowing how things work, being able to fix them when they break, and knowing how to create acceptable substitutes when they can't be repaired are not just hobbies but essential skills in a world growing ever more complex.

Check makezine.com/05/nff for related stories.

Tim O'Reilly (tim.oreilly.com) is founder and CEO of O'Reilly Media, Inc. See what's on the O'Reilly Radar at radar.oreilly.com.

BUILDING A SMARTER TO-DO LIST

PLANNING YOUR PROJECTS ONE BITE-SIZED PIECE AT A TIME.

By Merlin Mann and Danny O'Brien

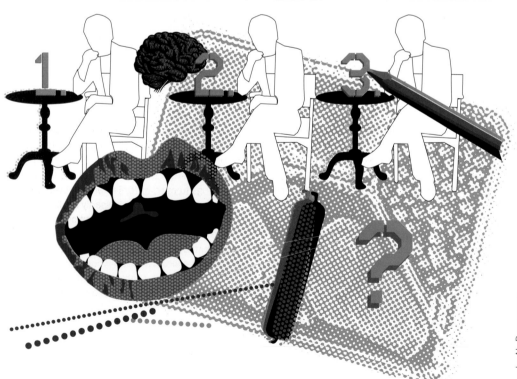

Illustration by Aly Rasmussen

WHILE YOU CAN ARGUE FOR THE flavor and approach to task management that best suits your style, it's hard to disparage the benefits that come from getting task commitments out of your brain and captured in a consistent location. Think of the to-do list as the evolving strategy for focusing your effort and attention in the immediate future.

Anatomy of a To-Do

The primary idea of a to-do is that it's a task that can and should be done — a point that might seem obvious until you start uncovering how many of the items on your to-do list may not belong there (or, conversely, how many uncaptured items do). The best and most useful to-dos share some common qualities: It's a physical action. It can be accomplished at a sitting. It supports valuable progress toward a recognized goal.

Glancing at your own to-do list, do you see any potential troublemakers? Notice any items that make you squeamish? Any "mystery meat" tasks that seem "un-doable" as is? Don't sweat it. We're going to have you shaped up in no time.

Let's Get Physical

Articulating your to-dos in terms of physical activity — even when they require only modest amounts of actual exertion — has a variety of benefits.

Most importantly, it ensures that you've thought through your task to a point where you can envision how it will need to be undertaken and what it will actually feel like once you're doing it; it's not just a bunch of words you've written on a page.

Framing your work in the physical world is easiest when you imagine what's being done, and the best trick here is to simply phrase your task in a form like: "verb the noun with the object." That means instead of reminding yourself with the mystery meat of "year-end report," you'd more accurately first start with "download Q3 spreadsheet from work server." And, instead of "get with Anil," you'd probably want to "email Anil on Monday to schedule monthly disco funk party." Get specific in whittling the task down to one activity that you can accomplish in ten minutes or less.

Now, Now, Now

Avoid the trap of littering your horizon with piles of crufty pseudo-tasks that can't actually be addressed (or, almost as often, can't be addressed *yet*). While you want to always stay aware of future obligations and the work that they are likely to generate, the to-do list is absolutely not the place to do it. Keep your to-do list a sacred tabernacle for current activity, and maintain longer-term task and support materials as well as appointments where they belong — in a project support folder and your calendar, respectively.

The trick is that most jobs can be made easier long before they're undertaken, simply by framing and naming them properly and in the right-sized units. As early as the capture and planning phases of building your to-do list, you hold the power and responsibility for defining your work properly. The last thing you want is to wonder whether you're doing the right thing at the right time.

> "Get specific in whittling a task down to one activity that you can accomplish completely at a sitting."

Why Am I Doing This Task?

When compiling a list of everything on your mind (and plate), it's crucial to unpack how each task you accept or assign to yourself will support your projects, your roles, and your goals. Before adding a new item, reflect on the value that each chunk of work brings to your world. Remember: Every minute you work on one task is necessarily a minute you can't work on another. Don't accept or assign crap. If it goes on your list, it's a commitment to seeing that task through to completion. Respect your time and attention enough to take your work seriously and plan it with care and forethought. Alpha geek status awaits once you can get your to-dos framed, flipped, and retired as efficiently as possible.

Note: This column originally appeared in an expanded version on Merlin Mann's site, 43Folders.com.

Learn how to reel in your mind at Danny O'Brien's lifehacks.com and Merlin Mann's 43folders.com.

Cory Doctorow

TRAITORS TO HISTORY

A COPYRIGHT-CONTROLLED MUSEUM IS A CRIME AGAINST HUMANKIND.

THIS PAST AUGUST, MY PARENTS CAME to visit me in London, and we had a grand sightseeing day that culminated in a visit to the Royal Greenwich Observatory, where clever makers in the 17th and 18th centuries made homebrew instruments to measure their physical environments, sensor arrays so clever that they enabled sailors anywhere in the world to divine their longitude. This was the scientific and military problem of the day, because sailors who can't calculate their longitude can't reliably cross oceans.

Today, the Greenwich Observatory is given over to a brilliant museum that tracks the history of this paleo-maker initiative. There are the most cunning brass clocks, handwritten logbooks in the crabbed script of Royal Astronomers centuries dead, and towering, proud telescopes built of wood and hand-polished optics.

And the first sign you see when you go through the door says "NO PICTURES." There's even a picture of a camera in a red circle with a slash through it.

The temple of measurement demands that it not be measured. The museum that is supposed to inspire a generation of scientists demands that the most basic tools of science — recording devices — be kept in our pockets.

I went and asked a curator why this was: was it because they're worried about flash pictures fading the exhibits? No — photons don't hurt old brass, no matter how precisely machined. The curator said that there's no photography allowed in the Greenwich Observatory because of copyright!

Copyright? How can that be? There's no such thing as a copyright on a clock. There's certainly no copyright in centuries-old logbooks; facts aren't copyrightable, and copyright runs out in substantially less than a couple hundred years — besides, the photographing of a single page out of a logbook

falls firmly in the realm of fair dealing, Britain's equivalent of the United States' fair use.

All right, the curator admitted. It's not really copyrighted per se, but we want to be the exclusive purveyors of photos and picture-postcards and so forth. What's more, some of the exhibits are on loan from third parties who made us promise to prohibit their being photographed.

There's no nice way to say this: a museum curator who takes this attitude to the exhibits in her charge is a traitor to history, to heritage, and to science. The point of a museum is to spread culture, not restrict it in order to run a penny-ante picture-postcard racket. A museum curator should not accept an unphotographable exhibit of historic interest any more than she should accept an exhibit

> "A wise steward of *David* would take every conceivable measure to see that this piece of world heritage was spread to every corner of the globe."

that comes with the condition that patrons who wish to see it must swear a loyalty oath, view a propaganda film, stand on one foot, or accept any other abridgment of their personal and cognitive liberty.

However, this attitude toward curatorship is fast becoming the norm, not the exception. For example, when the Florentine curators who control access to Michelangelo's *David* allowed Stanford to undertake a high-resolution 3D scan of the famous sculpture, it was on the condition that the resulting 3D files be restricted with Digital Rights Management technology to prevent their use in making copies of the *David* that were outside of Florentine control.

The *David* is not copyrighted. It is a centuries-old artifact that is part of the world's heritage. It does not "belong" to a museum any more than the

Mark Neel took this photo of Michelangelo's *David* and uploaded it to his Flickr.com account under a Creative Commons license.

Mona Lisa or the Elgin Marbles do; it is merely in the temporary stewardship of a museum that cares for it so that all of us may someday see this magnificent sculpture and understand a little more about what it means to be a member of the human race.

A wise steward of *David* would make it his business to see to it that every conceivable measure was taken so that this piece of world heritage is spread to every corner of the globe. The surest way to do that would be to use Stanford's high-resolution scan to print millions of these at the highest quality possible, distributing them to every school, park, museum, city hall, and public bath in the world. In an era of cheap, accurate 3D scanning and reproduction, there should be no person alive who lacks the opportunity to confront Michelangelo's masterpiece in living physical reality.

Instead, the stewards of *David* and innumerable other works of art take it as their duty to restrict access to their charges, all to sell a few tawdry postcards, to preserve their station and power as gatekeepers.

I've been to Florence. I've seen the *David*. Not only that, I've seen about a squillion copies of the *David*. There's a David on practically every street-corner in Florence, because for the past 500 years,

the way that budding Florentine sculptors learn their craft is by hacking out copies of the *David*.

Every creator stands on the shoulders of giants: so said Sir Isaac Newton. The sciences and the arts are built on copying, on observing, on measuring, on the public disclosure of facts and discoveries. The difference between alchemists (who didn't publish their results) and chemists (who did) is that every alchemist had to learn the hard way that drinking mercury was a bad idea, while chemists read their colleagues' papers on mercury, skirted danger, and devoted their energies to making genuinely useful new discoveries.

No curator of human knowledge has the moral right to restrict the recording of our shared history. The next time you find yourself at a museum — generally funded by your own tax dollars — where the rules say you have to keep your camera in your pocket, seek out the curator and ask him why he's turned traitor to history.

Cory Doctorow (craphound.com) works for the Electronic Frontier Foundation (eff.org) and co-edits boingboing.net. Doctorow's most recent novel is *Someone Comes to Town, Someone Leaves Town*, from Tor Books.

Graffiti Temple

Artist **David Best** is known for his gigantic temple installations at Burning Man in Nevada's Black Rock Desert; his latest was 115 feet tall, a quarter mile long, and visited by thousands. The temples — largely made out of scrap and the plywood cutouts left over from a local toy factory — are intended to be places of solace for people of all faiths. They're also a blank slate for everyone who comes to experience them. Visitors are encouraged to write or draw on exposed surfaces, so that the end result is a stunning collaborative statement of mourning.

This last summer, though, San Francisco Mayor Gavin Newsom invited Best to temporarily install a temple on Hayes Green in San Francisco, not far from City Hall and the opera house. While some might consider a plywood temple that actually *invites* graffiti to be a risky bet, Best points out that temporary art allows artists to bypass the red tape usually necessary to approve public art. This public work, however, was so well loved that it was up for almost six months, far longer than originally planned. By September, the surprisingly elegant plywood structure — part Thai temple, part Brussels lace — was covered with names, drawings, messages, stickers, photographs, and flowers as high as the arm could reach.

Best builds his structures with a part-volunteer crew, which includes everyone from a master architect to a person paralyzed except for the fingertips. He comes up with a drawing, and then he and his experienced crew try to turn it into reality. "There are no mistakes," says Best. "It's important to me to empower other people to work with the material." He has worked with thousands of volunteers over the years.

The Hayes Green Temple took 30 people three days to build. Still, despite the intricacy of the design and the shapes involved, the process is nothing complicated. According to Best, "It's real caveman stuff."

— *Arwen O'Reilly*

≫ **David Best Hayes Green Project:** blackrockarts.org/ david_best.html

Photography by Arwen O'Reilly

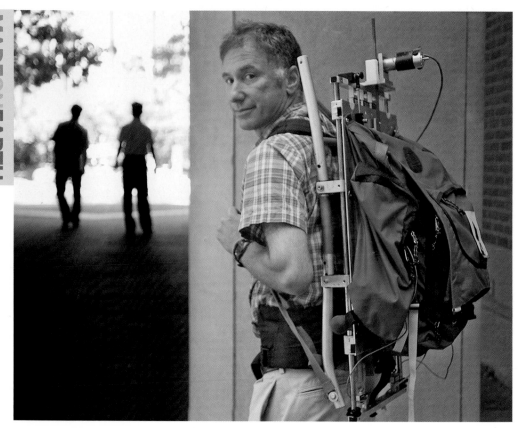

Power Pack

Finally, you can recharge your iPod with Clif bars.

When the military needed to recharge batteries on the move, they turned to University of Pennsylvania professor **Larry Rome**, an expert in muscle power and, it turns out, a capable inventor. His solution: the world's first electricity-generating backpack.

Rome, who studies fish muscles, says the idea struck him in a Navy meeting. U.S. troops were lugging 80-pound packs, including 20 pounds of batteries for high-tech gadgets. The brass wanted to use muscle power to generate electric power, but the best existing technology was shoe generators, straight out of *Get Smart*.

"I said, 'That's a terrible idea,'" recalls Rome. "The force of the heel strike is only over a couple millimeters. The right way became obvious: with every step, these guys are lifting 80 pounds 5 to 7 centimeters — that's potentially 36 watts of mechanical energy."

To turn his brainstorm into hardware, Rome grabbed an old external-frame backpack from college days and called his lab's "very fine machinist" Fred

Letterio. In their basement shop full of mills and lathes, the two added springs to suspend the cargo compartment from the pack frame. As the wearer's stride raises and lowers the pack, the load slides up and down, driving vertical rods to spin a geared DC servomotor up to 5,000 rpm to generate electricity.

With a 40-80 pound load, Rome's pack generates 7 watts, plenty of juice to simultaneously power a two-way radio, GPS receiver, and night vision goggles (or cellphone, PDA, digital camera, and iPod). The load can be locked for stability on sketchy terrain, and then unlocked to generate power again.

Ultimately, the generator pack (patent pending) will weigh just a couple pounds more than a regular backpack. Carrying it burns 3% more energy, but wearers say it's more comfortable, and the extra work costs only a couple of extra candy bars. ("Food is 100 times more efficient than batteries.") Green bonus: the technology could keep tons of toxic batteries out of landfills.　　　*—Keith Hammond*

≫**Lightning Packs:** lightningpacks.com

Flying Fruitcake

Back as far as 400 BC, catapults were used to hurl the vilest of projectiles: rocks, arrows, cow manure, diseased horse carcasses, barrels of venomous snakes, wasps' nests, the severed heads of captured soldiers. But perhaps the most notorious missile of all may be ... the fruitcake. At least that's what the citizenry of Manitou Springs, Colo., seems to think.

Each January, Manitou Springs residents gather for the city's annual Fruitcake Toss. The town's best and brightest combine their initiative and tinkering ability to build an array of mechanical and pneumatic-powered contrivances — including air cannons, catapults, and trebuchets — all designed with one thought in mind: to hurl unwanted fruitcakes out of town.

Toy designer **Dave Meyers** is the architect of the M-63 Fruitcake Remover. It's an industrial-grade, slingshot-style catapult powered by the ample potential energy stored in 36 feet of overextended ⅝-inch surgical rubber tubing. The machine has a yoke 10 feet across, and two large men are required to cock the spring. The trigger is similar to a sailplane hitch release, sailplaning being another of Meyers' interests.

Meyers built this machine large and strong because it has a big job to do. See, fruitcakes are hard to get rid of. The specific density of fruitcake is equivalent to mahogany. It's the one gift that the U.S. Postal Service has not found a way to damage in transit. And you simply cannot flush a fruitcake. Hence, the M-63.

"When I started, I essentially began doodling in AutoCAD," says Meyers. "I had a vision in my head for a slingshot, and this design just sort of developed as I stared at my computer screen."

One year, the M-63 launched more than 300 feet. Some of the larger air cannons present at the Fruitcake Toss hurl even farther. As fruitcake-slinging technology improves, the bar (and the fruitcake) gets ever higher.

—*William Gurstelle*

≫ **Fruitcake Toss:** makezine.com/go/fruitcake

LoCost Sports Car

Now midlife crises don't have to be quite as expensive. Ron Champion's cult classic, *Build Your Own Sports Car for as Little as £250*, has spawned a host of clubs devoted to doing just that. Slightly more complicated than your traditional kit car, the LoCost is assembled entirely from secondhand parts and raw materials (metal bar, aluminum sheeting), and requires quite a bit of welding.

You also need a donor car. "Basically you need an engine, a gearbox, and either a live rear axle or an entire independent rear axle unit — the rest is made up," says **Adam Streatfeild-Jones**, of Britain's LoCost Car Club. The chassis design is generally built from Champion's original drawings, but even a casual look at the images posted on enthusiasts' websites makes it clear that there's a range of vehicles. Some build road cars, for tooling around the countryside, and others concentrate their energy on racing cars. All the cars seem to be a labor of love.

In fact, LoCost car building became so popular that in 1999 the LoCost Formula race was started, run by the 750 Motor Club, with tight construction regulations to make sure that the winners are not just the ones with the deepest pockets. In fact, the founders intended it to be a way to encourage younger drivers and builders, and entries include cars submitted by colleges and groups as well as thrifty individuals.

According to one website, a car has been built for as little as £47.50, but Streatfeild-Jones admits that "most builders want their car to look good, and thus they go for expensive paint jobs, nice wheels, and so on." Even with all the trimmings, it's still a far cry from the cash needed to buy a sports car hot off the assembly line. So don't let that $360 burn a hole in your pocket.

—Arwen O'Reilly

≫ **The LoCost Car Club:** locostcarclub.co.uk

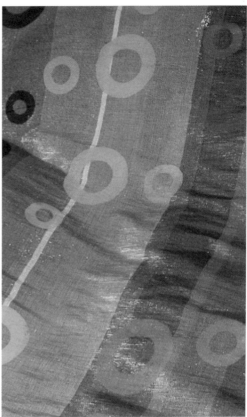

Photography by Monica Mills (left) and Alyce Santoro (right)

Sonic Fabric

Throughout history, music has played an important role in our social, political, and religious rituals. It's been used to express the most elusive of emotions and situations: love and loss, conquest and oppression, joy and mourning. You name it, there's a piece of music that will fit it. And now, the music can fit you: **Alyce Santoro** has created a series of products that allow you to wear your music and listen to your clothes.

As a child, Santoro would watch the tell tales, made from cassette tape, of her family's sailboat, and imagine them playing their song to the breeze. Later, she learned about Tibetan prayer flags, whose sacred mantras are released by the wind and carried out into the world. The two seemed to fit together.

Using cassette tape recorded with music and sounds significant in her life, the New York-based artist started making fabric. She had planned to produce enough material to make only a few prayer flags, but when her friends saw her knitting and crocheting

tape, they too became inspired. With the added input, the project took on a life, or lives, of its own.

Her friends suggested various techniques of weaving and decorating the cloth, but friend Joe Ball made one of the most significant contributions. He suggested they might be able to listen to the tape. Until then, the music was purely symbolic. Sure, Santoro knew it was there and what it represented to her, but her creations remained silent to others.

"It hadn't occurred to me that it might actually be possible to hear them," she says. "And sure enough, by simply retrofitting an old Walkman by mounting the head on the outside, Ball created the first Sonic Fabric reading apparatus."

Santoro even crafted a dress for Jon Fishman, drummer from the band Phish. The Sonic Fabric was woven using tape culled from Fishman's extensive collection, and the dress was "played" onstage in Las Vegas, using specially designed tape-head gloves during a 2004 Phish show in.

—*Josie Moores*

>> **Sonic Fabric:** sonicfabric.com

Spatial Education

Video game composers have to be crafty, because every project is a new world with new rules. **George** "the Fat Man" **Sanger**, who's been at the top of the craft since the Mattel Intellivision ruled the Earth, is crafty indeed. His homemade Leslie speaker — which he first built out of a punctured paint can and record player — taps a polyethylene drainage device to get a sound he describes as "one of the most organic, spacious, and beautiful there is."

Invented by Don Leslie (1911–2004), the Leslie rotating speaker defined the sound of the Hammond organ, becoming a staple of rock, jazz, gospel, country, and pop music. It's frequently used on guitars as well. As the speaker spins toward and away from the listener, the Doppler effect alters the sound's pitch and timbre, producing a rich vibrato. And the way the projected sound sweeps around the room? "Man, that is good medicine for the soul," says Sanger.

To make his desktop Leslie, Sanger used a garage-sale turntable ($5), a 4x1-foot wire shelf ($4), and a double-outlet catch basin ($7). He then added plastic cups to extend and focus the sound. Inserting a CD-R spindle loaded with unwanted discs into the basin's bottom provided a solid base and allowed the cups to clear the tone arm.

By increasing the diameter of the rotor, the cups also heighten the Doppler effect, because the output ports travel faster. Sanger found that one foot is the optimal diameter at standard turntable speeds. The amp he used is a Vox Brian May Special (about $149 new), which has a big sound and a sealed back, making it very directional. Its controls are on the top rear, which puts them in perfect operating position when the amp is face down.

Next up: Sanger is currently plotting how to make a Leslie speaker out of a ceiling fan.

—*David Battino*

≫ **The Fat Man:** fatman.com

▣ **Hear audio examples at** makezine.com/05/made.

Photograph by George Sanger

Photograph courtesy of Phoneticontrol and Shawnimal

Huggable Tee

Making something out of T-shirt material can mean different things to different people. For **Phoneticontrol** and **Shawnimal**, it meant designing Pulse Width (PW), a "happy little 'bot" filled with "stuffing and love. Mostly stuffing."

The bug-eyed, blue collaboration was created as an entry in I Love Your T-Shirt's (iloveyourtshirt.com) competition for *something* made out of T-shirt material. Named after a term used in analog synthesizers, Pulse Width was designed in two dimensions by the pizza-loving **Eric Broers** (Phoneticontrol), an illustrator and designer. "We both needed to change our thinking a little to bring him to life," Broers says. "It was an interesting challenge. I designed the PW, and Shawnimal sculpted the crap out of it."

Shawn Smith (Shawnimal), a multi-talented artist who paints, sculpts, draws, and creates plush creatures, brought the vision into the third dimension. "The challenge was getting Eric's style to come through in plush form," he explains. "Somehow, maybe the planets aligned or some-thing, but I think we nailed it."

The two have worked together in the past and had a blast, so when Phoneticontrol found out about the contest, he immediately thought of working with Shawnimal to create a plush. Once it was created, the duo decided that Pulse Width needed to get airborne. After a little old-fashioned tossing into the air and rapid-fire photography, PW was captured airborne on film above Logan Square in Chicago, where they both live.

Smith, who sells both big and pocket-size plush ("because small equals cute"), ends on a philo-sophical note: "All I can say about Pulse Width is frumpy, huggable robots rule."

—*Shawn Connally*

≫**View photos:** phoneticontrol.com
shawnimals.com/projects.php
studio606.com

Bruce Sterling

HANDS ON:
MAKING A MOBILE

STAYING POISED AT THE FERTILE EDGE OF CHAOS, WHILE THE CLOCK TICKS.

I'M IN RESIDENCY AT ART CENTER COLLEGE of Design in Pasadena, Calif., and it didn't take me long to figure out that most everybody here makes stuff.

Here at design school, it's a demo-or-die situation. For instance, everybody in my "Ecology of Things" class is busily making demos and prototypes for highly distributed, embeddable computer chips. These microchips are supplied to the students courtesy of the research and design wing of Sun Microsystems.

Since I'm an author and have spent my entire career blabbing, I decided my design students would surely benefit by seeing me build something myself.

The prospect of ubiquitous, ad-hoc'ed, networked, pervasive, teensy, radio-frequency, spooky, action-at-a-distance, internet-of-things-ish, Java-swapping, motorizable chips is not just talky, but a little daunting. Luckily, there's already a sweet, artsy, familiar, and highly popular version of a working "ecology of things." It's from the pre-digital age. It's called a "mobile."

Alexander Calder (1898-1976), that Modernist artist with a mechanical engineering degree, invented the mobile in the early 1930s. The materials for mobiles — steel wire, scrap aluminum, and paint — are cheap and easy to find. A vintage Calder mobile goes for upwards of half a million bucks at art auctions now, but Sandy Calder used to be able to whip one out in a day. So how tough could it be to just make one?

Equipped with hardware-store wire and cheap Chinese hand tools from the local Target store, I soon discovered that it's, in fact, dead easy to make a mobile. A mobile is simply a cascade of levers, a multilevel, energy-trophic ecology of levered elements, if you will. The key to success is the relationship between elements. If the links are

made too tight and rigid, then the mobile has a lifeless, clockwork, phony feel, much like a Soviet bureaucracy. If the links are too loose, then it gets all corrupt and herky-jerky; its extreme fringes start clattering around in violent incompetence, much like the Bush administration.

A successful mobile artfully combines central control and local initiative, artfully poised at the fertile edge of chaos. Ideally, the thing should dance.

My own model mobile (after many returns to the drawing board) dances just fine now. Unfortunately, since I lack physical skill at artfully working steel wire,

> "A vintage Calder mobile goes for upwards of half a million bucks but Sandy Calder used to be able to whip one out in a day."

it's ugly; it looks like it was dropped off a ten-story building. That doesn't matter, though, since the original model, made from paper plates, was merely the working prototype for a far more ambitious effort. My completed mobile is supposed to have a motor on board, and an embedded microchip to assure interaction with passing humans. It's going to be about 40 feet across.

This entails scaling up my model by a factor of ten. Scaling up chaos by a factor of ten is a rather nonlinear effort. The sweet and subtle balancing forces within my tiny mobile maquette are packing some serious heft now. In my misspent youth, I used to stretch barbed wire and string Texan cattle

Photography by Bruce Sterling

fences. Every once in a while, under pulley tension, a strand would pop and lash free, wickedly eager to scar and/or blind. Then you'd learn instantly that barbed wire is a torture device with extensive military applications.

As I wrestle with my wobbling polyvinyl pipes and dangling chunks of illustration board, that early lesson is returning with redoubled force. I now finally and fully understand the wisdom of a good design-school education. You can lie to the board of directors, but folks, you just can't lie to steel wire.

I'd love to rattle on about this subject, because the many weird, sexy applications for digitally interactive mobiles are so intensely novelistic and science-fictional. But I've got a deadline and a budget to meet. The clock is ticking and I'm running out of cash. I'm learning a lot really fast. So I'd like to keep talking here, but, well, I gotta go.

Bruce Sterling (bruce@well.com) is a science fiction writer and part-time design professor.

Maker

Woody's World

A Q&A with the inventor of a pirate-scaring noisemaker, a helicopter for every garage, and a way to hack gravity itself.

INTERVIEW BY WILLIAM LIDWELL
PHOTOGRAPHY BY JENNY PFEIFFER

Woody Norris is standing 50 feet away from me, pointing a menacing-looking sci-fi contraption at my head. We're in the back lot of his American Technology Corporation in San Diego, Calif. Woody's device is called the LRAD (for Long Range Acoustic Device). It looks something like a big spotlight, the main difference being that it projects sound instead of light. Woody yells to me, "We are going to start with some music. This is really cool!"

Now hear this: Elwood "Woody" Norris'
Long Range Acoustic Device was developed
for the U.S. military to send a painful blast of
noise at small boat crews attempting to get
near U.S. vessels.

He fiddles with the controls and I hear classical music as clear as if I were next to the speaker. I smile and nod back. Woody grabs a microphone and speaks through the LRAD, "Now I'll play some pre-recorded commands being used by our troops in Iraq." I can hear him clearly. I begin to hear a series of commands like "Halt" and "Put your arms over your head" quickly followed by their Arabic equivalents. Again, I nod. Then I see Woody and his prototyping guru, Jeff Belka, chatting back and forth, nodding, and smiling. They are up to something. Woody looks my direction and all I see is a big set of teeth slowly revealed by a widening smile.

He turns a control on the LRAD and a 150db siren screams out of the device — my hands instinctively shoot up to cover my ears. The siren is so loud I can't think or do anything. I am paralyzed. The siren lasts for only a couple of seconds before Woody turns it down. I start to put my hands down and he cranks it. My hands shoot back up and I stand there at his mercy. We do this dance a few times until Woody is convinced that I get it. The LRAD is cool technology. It has all manner of communication, crowd control, and acoustic weapon potential. Before I can clear my head and convert my thoughts to words, Woody starts to reach for the LRAD again. "It is cool! It is cool!" I shout. He gives me an approving look and motions me in. "Yeah, cool," he says. "Now we can continue."

A recent winner of the Lemelson-MIT Award for Invention and holder of more than 40 patents, Woody Norris is the classic American success story. Born of modest means and largely self-educated, Woody is the founder of a number of companies based on his inventions, ranging from personal flying aircraft to a variety of revolutionary acoustic technologies. As inventors go, he is a rare breed that enjoys selling as much as inventing — an unusual blend of P.T. Barnum and Thomas Edison.

Woody's workshop is within eyeshot of his huge and opulent home on an adjacent hilltop. As we go inside, my eyes are immediately drawn to the notes and diagrams on his whiteboard. Intriguing terms like "plasma antenna" and "optical cobweb" fill the board. On the workbench, I spot some very strange-looking speakers, a gas chromatograph, and circuit boards that he appears to be reverse engineering. Seeing this array of gadgetry leads to my first question ...

When did you discover that you wanted to be an inventor?

My whole life I have been dragged along by circumstance. I became an inventor by accident. Around 1960, I was reading a magazine that had an article about a new kind of electric shaver that ionized off your whiskers — it had drawings and everything. The article intrigued me. The further I read, the more incredulous I became, and sure enough, at the end it said, "April fool." Here is the part that changed the rest of my life. To the side of the article, there was a small box, outlined in black and titled, "Editor's Note: Submissions Wanted." It was a contest announcement inviting readers to write next year's article. The winner would receive $250. I was only making about $400 a month at the time, so that was a lot of money to me. I dropped what I was doing, dreamed up a new invention, and wrote the article.

My invention idea was a record player tone arm that made a straight line across the record instead of pivoting at an arc. Since that is how they cut records, it seemed reasonable to me that a straight-line playback arm would be superior. I was literally in the process of licking the envelope to submit my article when I stopped myself. I decided that I should call some hi-fi stores pretending that this was the real deal and see if they would believe it — the thought being that if it could fool them, then it could definitely fool the readers of the magazine. So I called up every retailer in the Seattle area and not only did they believe it, they were excited about it. At that point, I decided I better build the thing instead of submitting the article. And that set me on the path to inventing things.

Of your many inventions, I am most intrigued by the HyperSonic Sound technology. How does it work?

HyperSonic Sound is my landmark invention. It allows you to project sound like a very focused beam of light. With a normal speaker, sound propagates outward as waves in the air. If I point a normal speaker away from you, you still can hear the sound. With a HyperSonic Sound speaker, if I point the speaker away from you, you won't hear anything. How does it work? Ultrasonic frequencies are very directional because their wavelengths are so short — the higher the frequency, the more like a beam they become. So I figured out a way to increase the frequency of standard audio content to ultrasonic levels. If you do this in a particular way, the ultrasonic signal demodulates in the air and turns into audible sound. So instead of emitting audible sound from a single source like a speaker, the sound is actually made in the air at an infinite number of points along this ultrasonic column. It's like having a billion little speakers lined up and pointing in the same direction. It is very cool — like magic.

Can you describe your process of invention?

I largely invent by analogy. I find things that are successful in one environment and then think about parallels in other environments where people have not yet discovered or exploited the concept. For example, I came up with the idea for HyperSonic Sound by analogy. I was watching television one day and I noticed that the logo on the TV had three colors: red, green, and blue. These are the colors that are squirting onto the screen from the back of the picture tube from which millions of colors are created. By mixing just a few colors, you can get millions of colors. I thought to myself that this was pretty cool. And then I thought about electronic circuits, where there are components called mixers that allow you to mix a couple of frequencies together to get new ones. So it occurred to me that if you can mix energy optically and electronically, why couldn't you mix sound energy in a similar way? Then I started thinking about the possibility of mixing sound that is inaudible to people — in this case, ultrasonic sound — in a way that would result in audible frequencies. And that is where the HyperSonic Sound idea came from. I solved the artificial hip problem in a similar way.

The artificial hip problem?

When someone gets an artificial hip, it can fall out of socket quite easily. It takes many months of conscious effort for patients to learn how to do things in a way that prevents the artificial hip from disengaging. When one disengages, it is very painful and costs thousands of dollars to reset. Anyway, I was approached by a group of doctors who wanted a way to detect when an artificial hip started to separate and then set off an alarm. In this way, people could stop what they were doing and tense up their muscles to prevent the hip from disengaging. Additionally, they didn't want any active electronics or power sources to be inside the body.

Now, around the time they contacted me, I had been playing around with a guitar. I noticed that when one string was tuned to the same note as another string, plucking one vibrated them both. This is sympathetic vibration. So I started thinking about the string I plucked as a radio transmitter sending out a signal and the receiver as the other string that is tuned to the same frequency. The second string did not start vibrating for free — it takes energy to get it to move. It then occurred to me that I could make a transmitter that sends a signal into the body, and if there was a thing inside the body that was tuned to the same frequency, then we could use the change in frequency as a trigger. So I embedded a resonant circuit — which is simply a coil of wire — in the artificial hip that was tuned to the same frequency as an externally worn transmitter. If the hip started separating, the

**1. American Technology, Norris' company, donated several MRAD "sonic lasers" for Hurricane Katrina crowd control and rescue operations.
2. A component of an artificial hip alarm, which warns patients that their implant is about to dislocate.
3. A HyperSonic Sound speaker.**

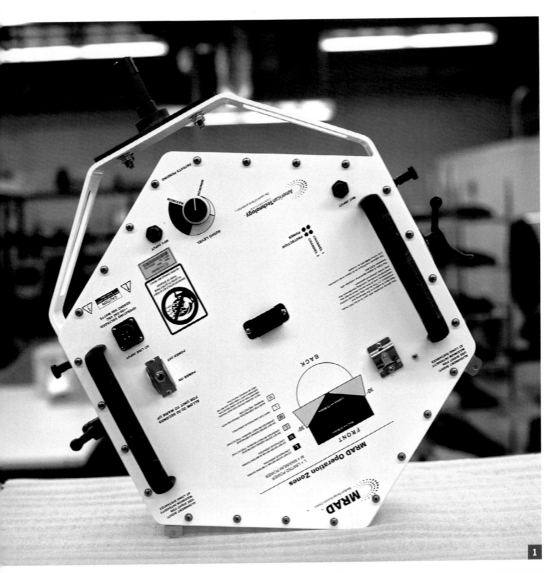

frequency would shift, and that shift would trigger an alarm. Sure enough, it worked.

The AirScooter is a bit of a departure from your other inventions. What is the story behind it?

My business partner and I began discussing how people love to fly. So we decided to invent a flying machine. I then made a list of all the things I didn't like about airplanes: I don't like to have to be going 100 mph to stay in the air, I don't like having to drive to the airport, etc. I then made a list of all the things I didn't like about helicopters: I don't like how hard they are to fly, I don't like the cost to get licensed, etc. Then we simply went about coming up with a product that met all of these criteria.

We knew people needed to be able to take off and land anywhere, so we went with a rotor-based craft. The tail rotor on helicopters drains about 20% of the power without adding lift, so we replaced it with a second rotor that counter-rotates. This eliminated the wasted power and the gyroscopic effects. We knew the controls had to be easy to use, so we eliminated the foot pedals and yoke and went with a handlebar. Push, pull, or turn it the direction you want to go and you go there.

We knew it had to be low-maintenance. The ratio of fly time to maintenance for a helicopter is about 1:1, which is not acceptable to most people. If we could make it light enough, we could eliminate a lot of the rotor complexity (and corresponding maintenance) and control lift through rotor speed. Additionally, if we could get the craft under 254 pounds, it would qualify as an ultralight and be exempt from licensing requirements. So we reduced the weight, primarily by inventing a new high-power, lightweight, four-stroke engine. This allowed us to get the weight down and reduce the rotor complexity. Now a pilot can use a simple handlebar throttle to go up and down. It has taken us several years to work out all of the kinks, but for the price of an expensive car ($47,000), anyone can now buy and safely fly their own AirScooter out of their backyards. Pretty cool!

Rumor has it that you are working on some inventions that involve gravity. Anything you can share?

I am fortunate that I am now in a financial position that allows me to pursue big problems. So for the last several years, I have tried to deliberately invent over my head. When I can't get the answers I need for problems I'm working on, I find professors who specialize in the areas that I am working on and then pay them to conduct the research or write me papers. This allows me to get experts in niche areas to evaluate my ideas and tell me if they are smart or stupid.

So with that context, let me say that nobody really understands gravity. What we do know is that it has something to do with mass, and it is the result of something going on in the nucleus of the atom. I think I know what that "something" is. I might be the only person in the world who knows what that "something" is. I know that sounds brazen, but I believe it's true. I have been working on a variety of experiments, over the course of years, to test my theory in a way that yields clear, unambiguous results so that it can be independently reproduced — I don't want a Pons and Fleischmann situation (*see MAKE, Volume 03, "A Fusion Reactor for the Rest of Us," page 25*).

Based on the current rate of progress, I am confident that within ten years we will understand the mechanism of gravity and be able to influence it in specific ways. I realize that most scientists reading this will think I am nuts, but I am not risking anybody's money or reputation but my own. So if I am wrong, no big deal. If I am right, the rewards for everyone will be terrific.

Read more of William Lidwell's interview with Woody Norris at makezine.com/05/interview.

Racks of HyperSonic Sound speakers ready to ship. Customers include Disney World and supermarkets, who use them as "audio spotlights."

What Woody's Wrought

Norris' key inventions, from audio spotlights to flying cars.

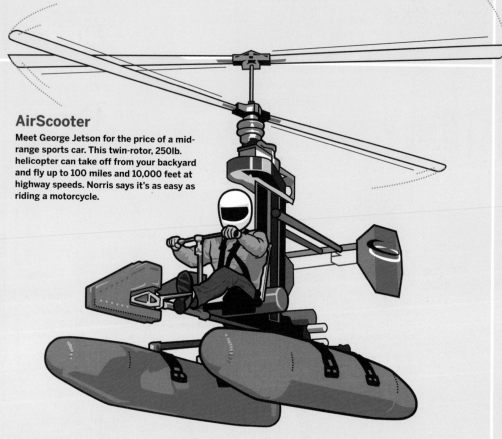

AirScooter

Meet George Jetson for the price of a mid-range sports car. This twin-rotor, 250lb. helicopter can take off from your backyard and fly up to 100 miles and 10,000 feet at highway speeds. Norris says it's as easy as riding a motorcycle.

HyperSonic Sound System

The first major breakthrough in acoustics since the loudspeaker was invented 80 years ago. HyperSonic Sound speakers send out two different ultrasonic signals. When they arrive at their intended destination (any object within 300 feet), they combine to form audible sound.

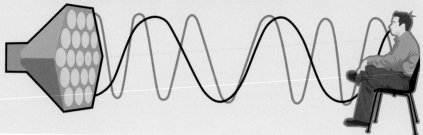

Illustrations by Dustin Amery Hostetler / UPSO.org

Long Range Acoustic Device

Related to HSS, the LRAD can project audio in a tightly focused sound beam over 500 yards at insane volumes. Good for any situation requiring long-distance communication. Want to disperse a drunken riot after a football game? Point one of these at the mob, crank it to 150db, and watch them scatter. Norris' LRAD was recently used by the crew of a cruise ship in Africa to scare off pirates trying to board the vessel.

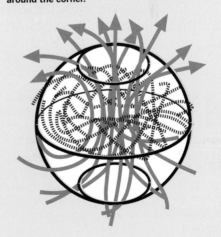

Gravity Machine

Gravity shield? Gravity distortion device? Woody has been conducting experiments and working with experts. He is not talking details, but stay tuned — he says something big is just around the corner.

Ear Radio

At less than ¼" thick and ¼ ounce, this tiny FM radio caught the attention of NASA engineers and ultimately led to the first hands-free earset, which integrated the microphone and speaker in the earpiece and used bones in the head to conduct sound.

Art in Living

N55, a conceptual group from Denmark, creates functional art for living, breathing, and growing.
By Bruce Stewart

MAKING PUBLIC SPACES INHABITABLE and useful to anybody is the goal of N55, an artistic effort based in Denmark that has created functional art for over a decade. N55 creates inexpensive, modular, portable structures. Installations have included floating platforms and living structures, urban-farming plant modules and floating fish farms, and even a mobile bar. At the core of N55's values is the idea of art as part of everyday life.

Originally a collective of a dozen artists living, working, and exhibiting together, N55 was later pared down to the husband-wife team of Ion Sørvin and Ingvil Hareide Aarbakke. With the passing away of his wife last year, N55 is now carried on solely by Sørvin.

He is currently working on the first module for a new system called *Micro Dwellings*, a low-cost, easy-to-build, amphibious modular house designed to function on both land and water (floating or submerged). The modular *Spaceframe* is a reconfigurable, easy-to-build, low-maintenance living space structure. Other projects include a *Floating Platform* that can support these *Spaceframe* buildings, and a *Snail Shell System* that resembles a giant plastic shell-like wheel and lets people "move around, change their whereabouts, and live in various environments." Various cramped environments anyway. Many of N55's projects are about demonstrating the minimum space and materials needed for basic survival and producing them easily and cheaply.

N55's projects range from the practical to the political. Besides dwellings, food-production systems, and portable soil factories, there is a language education kit for illegal immigrants in the works, and a "protest" rocket that can distribute various things over a vast area from high altitudes (for example, printed matter or seeds).

One of N55's most popular and long-standing projects is *Land*, which encourages people to donate property for public use. Parcels of land have been contributed to this effort in the United States, Switzerland, Denmark, the Netherlands, France, Norway, and Sweden. The land is maintained by whoever is currently using and inhabiting it.

N55 started out as an artist's group in 1994, when they were 12 members living together in an apartment in the center of Copenhagen, trying to "rebuild the city from within." The group took its name from its address at Nørre Farimagsgade 55. Sørvin and Aarbakke continued to make use of this space as the launching pad for their functional art

The Snail Shell System **(previous page) is a low-cost way for people to move around and live in various environments. Pictured is the late Ingvil Hareide Aarbakke, one of the founders of N55. The** Spaceframe **(above) is low-cost and movable; N55 envisions it as a living space for three to four people. "It demands practically no maintenance."**

projects. Financed primarily by exhibitions, grants, and educational work, Sørvin considers N55 non-commercial, and creates and documents all the projects expressly so they can be freely duplicated and refined by others around the world. You could say N55 makes open source domestic technology.

N55 publishes manuals for all of its projects online (N55.dk), including detailed instructions and photos, so others can easily implement the ideas. There is also an N55 book that includes most of these manuals (available as a free download).

Bruce Stewart has covered technology issues for various publications for over a decade and is the editorial director of O'Reilly Media's online publishing group.

C.S. and the City

Intel's Eric Paulos and the edges of urban computing.

By David Pescovitz

There's a bomb inside Eric Paulos' storage space. It's not your typical homebrew explosive, though. This is an information bomb. If you happen to walk by when Paulos fires it off, you may not even notice right away. But your wristwatch will probably stop. Your credit cards certainly will no longer be readable. Your cellphone will instantly become a paperweight. Carrying a laptop computer? Hopefully, you have a recent backup. With the press of a single button, the I-Bomb unleashes a powerful electromagnetic pulse that kills electronic devices and corrupts all storage media within several meters.

Paulos has demonstrated the I-Bomb at places like the San Francisco Museum of Modern Art, but he'd be ill-advised to show it off at work. Not to say that his colleagues wouldn't appreciate it. In fact, projects as unusual and provocative as the I-Bomb landed Paulos his job in the first place. Still, he works for a company where fried chips are particularly distasteful.

A BIZARRE MACHINE LABLET

Paulos is a computer scientist at Intel Research Berkeley, a corporate "lablet" in the penthouse of a downtown office building where a dozen or so scientists explore the edges of computing and networking technology. When Paulos joined the lab several years ago, the first thing he did was set up a machine shop with lathes, scopes, and assorted prototyping tools. The laser cutter took a bit more wrangling but is finally due to arrive this week. A computer science Ph.D., Paulos can hack C++ with the best of them, but he's really a lifelong maker who made a career out of building bizarre machines.

As a graduate student, he designed groundbreaking telerobots that enabled people to physically explore remote spaces over the internet. His work at UC Berkeley bled into years of collaboration with infamous San Francisco machine performance group Survival Research Laboratories. In 1997, Paulos and SRL director Mark Pauline invited anonymous web users from around the globe to remotely aim and fire a massive air launcher loaded with concrete-filled soda cans. It was the first time lethal machinery was operated over the internet. Paulos' own tech-art collective, Experimental Interaction Unit, has exhibited around the world.

At Intel Research, Paulos directs the Urban Atmospheres program, a group exploring technology in the cityscape, from Bluetooth-enabled phones to ad hoc networks of tiny sensors. While many companies are developing urban computing applications — from location-enabled restaurant recommendation systems to real-world buddy lists — Paulos says he prefers to look between the cracks in the asphalt for research topics. After all, entire conferences are already devoted to the likes of location-enabled services and Geoweb technology.

"There are certainly moments in life for productivity and efficiency, but there are also moments for wonderment and reflection," he says. "I'd like to use technology to celebrate non-places, non-events, non-activities that actually matter a great deal to the emotional experience of urban life."

ONE MAN'S TRASH IS ANOTHER MAN'S RESEARCH PROJECT

Urban Atmospheres aims to illuminate the most subtle characteristics of city living, like the relationships you have with people you see every morning at the train station but never acknowledge, the invisible cellular infrastructure you move across, even the

TRASHCAN

For Paulos, every city is an ideal laboratory to test what he calls "objects of wonderment."

Photography by Emily Nathan

Left: Jabberwocky is a mobile phone application that detects "familiar strangers" nearby. To avoid privacy issues, the color and motion of the blocks on the display provide information about crowds rather than specific individuals.

Right: One prototype for the Connexus project is a "friendship bracelet" containing superbright LEDs that display a range of color and tactile outputs based on input from another bracelet wearer. Another version mimics the form factor of a wristwatch. Surrounding the Connexus devices are several "motes," coin-sized wireless sensors. The matchbooks were part of an experiment about anonymous text messaging.

public trash cans you walk by on the way to the office.

Last year, Paulos and Tom Jenkins, a student at the Royal College of Art, spent countless hours stalking a single garbage can in San Francisco's Financial District. After their field study, they built what Paulos half-jokingly refers to as "the most expensive trash can in the world" and set it on the corner outside the lab. Called Jetsam, the can is augmented with a variety of hidden sensors, processors, and a video projector. An infrared switch detects when refuse has been tossed into the can (or grabbed out of it), and snaps a digital picture. A highly sensitive scale then weighs the item and a laptop PC categorizes each bit of refuse by time and size. Meanwhile, an evolving visualization of the garbage photos and data is projected out of the can's opening, creating a rotating galaxy of garbage on the sidewalk.

"Archaeologists dig through trash to learn history, so we looked at what trash can tell us about urban living," Paulos says. "Is there a lunch binge at noon or a coffee craze at 3:15? If there is a continuous stream of lottery ticket stubs, does that mean the local denizens are risk takers? Instead of hiding trash, we changed it into something that people interact with."

And they did. A young graffiti artist tagged a piece of paper and then tossed it into the can, projecting his name onto the sidewalk. Later, a businessman exiting a photo developing shop tossed his unwanted double prints into Jetsam's mouth, creating an instant photo album on the pavement.

"It's not as if we think Intel should be developing trash cans," Paulos explains. "But we want to propose research ideas that might feel a little awkward, because it's from those unusual vantage points that we may see other interesting possibilities."

> "Making things is much harder than programming because there's no quick Control-Z undo."

KEEPING TABS ON YOUR FAMILIAR STRANGERS

Indeed, Paulos charts urban territories and dynamics that are invisible, ignored, or taken for granted. Sashay is a mobile phone application he wrote that maps your path through the various cells in a city's mobile phone network. The point of Sashay, Paulos says, "isn't to help you get places, but rather cause you to reflect on how you move." Another project for mobile phones, called Jabberwocky, extended the Familiar Stranger experiments conducted in the early 1970s by social psychologist Stanley Milgram. A familiar stranger is someone who you may see repeatedly, at the bus stop or cafe, for instance, but never interact with. You mutually agree to ignore each other. The Jabberwocky application grabs the

unique Bluetooth identifier of the mobile phones around you, creating a log of people you've previously encountered, even if you've never talked with them. Later, a quick glance at your handset reveals if any of your Familiar Strangers are nearby.

"You might not want to make friends with everyone you see regularly, but these people color your city," Paulos says. "Jabberwocky enables you to get a sense of place based on who is there with you."

A previous project, Connexus, highlights another form of human interaction that's rarely discussed: nonverbal communication like the friendly smile across the room or a reassuring pat on the back. Connexus is a bracelet meant to contain a wireless radio, sensors, and actuators to facilitate nonverbal communication at a distance. Imagine a couple both wearing Connexus devices and perhaps miles apart. As the woman taps on her bracelet, her husband notices his bracelet glowing softly, not unlike a vintage mood ring. He responds by caressing the bracelet, an input that manifests itself as a gentle heating sensation on the other end. Paulos never managed to stuff all of the Connexus components into as small a wearable device as he'd like, but he says that prototyping is essential to communicating his research goals to others.

ENVIRONMENTAL RINGTONE SOUVENIRS

Right now, he's fabricating a device that he hopes

will add a layer of audio intrigue to the urban atmosphere. Installed, say, in a public commons area, the small object will grab the Bluetooth identification numbers from the mobile devices of passersby. That data will then be translated into a unique ringtone, creating a souvenir of that location tied to a moment in time and the specific group of people nearby.

The device itself is only one part of the project. Paulos also plans to release a free developer's toolkit to inspire others to create and deploy their own interactive objects into the urban atmosphere. For Paulos, every city is an ideal laboratory to test what he calls "objects of wonderment."

"Making things is much harder than programming because there's no quick Control-Z undo," he says. "But if you're explaining your research to someone and can't present them with an object, they're likely going to default to talking about things they know. On the other hand, if they can interact with an object outside their realm of understanding, they're usually more willing to come with you into the conversation."

David Pescovitz, MAKE's special projects editor, is also co-editor of BoingBoing.net and an affiliate researcher with the Institute for the Future.

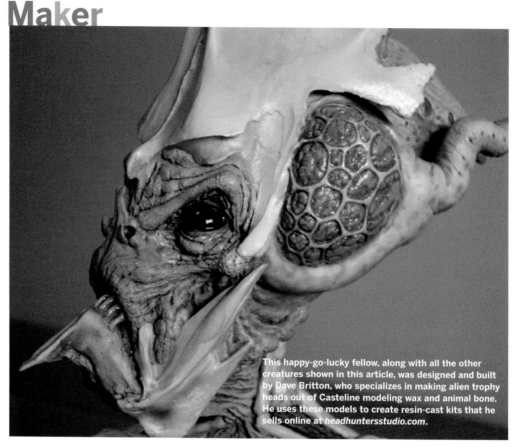

This happy-go-lucky fellow, along with all the other creatures shown in this article, was designed and built by Dave Britton, who specializes in making alien trophy heads out of Casteline modeling wax and animal bone. He uses these models to create resin-cast kits that he sells online at *headhuntersstudio.com*.

It Came from My Garage!

Model kit makers bring B-movie monsters to your home.

By Gareth Branwyn

MANY MAKERS LIKELY GOT THEIR first bug for building by putting together styrene plastic vehicle and monster model kits as kids. Until the late 1980s, the barriers to entry for actually fabricating — as opposed to assembling — such kits meant that only commercial concerns could afford to do so. But as new, easy-to-use materials for sculpting, molding, and casting became cheaper and more readily available, kit creation became a possibility for just about anyone with a lick of sculpting talent, a few bucks in the old pocket, and a world of patience. Combine all this with the power of desktop publishing, the internet, and e-commerce, and the "garage kit" modeling movement was born.

Until the emergence of garage modeling, builders were limited to commercially available model kits, mainly military hardware, cars, planes, spaceships, and mainstream movie monsters. But DIY kit creators cast their nets much wider. Today, a model

Photography by Dave Britton

enthusiast can buy kits of characters from David Lynch films, low-budget horror flicks, underground comic books, and spaceships from just about every cult sci-fi classic. And then there are the crazed, whimsical kits, models of aliens, monsters, pinup babes, and mad machines, rendered directly from the fevered imaginations of their sculptors.

Terry J. Webb, editor of *Amazing Figure Modeler* (amazingmodeler.com), one of the premier garage kit magazines, is also the author of three books that chronicle the development of the do-it-yourself kit scene, from its beginning in Japan in the 1980s to the present. With titles like *The Garage Kit That Ate My Wallet*, *Son of the Garage Kit That Ate My Wallet*, and *Revenge of the Garage Kit That Ate My Wallet*, these books say a lot about the humor that underlies much of this movement, the keen interest in B-movie subject matter, and the expense of these small-run resin- and vinyl-cast kits, which can cost from $60 to several hundred dollars. "A lot of what drives this hobby," says Webb, "is a love of sci-fi and fantasy movies and television. Many people get into kit making because they love a movie or show so much, they want to create a little 3D monument to it that other people can enjoy."

Many of the model makers behind garage kits are involved in the movie business, often special F/X techs, make-up artists, and prop masters. It is through these creative professions that they get their sculpting talent and their familiarity with molding and casting techniques. Dave Britton of Headhunters Studio (headhuntersstudio.com) is cut from this mold. He worked at several Hollywood F/X shops in the 90s, but soon found himself in Portland, Ore., where he started to make and sell latex monster masks. It was through his mask business that he discovered the kit-making scene and soon began producing figure-modeling kits.

Unlike many sculptors and casters who make direct copies of movie, TV, or sci-fi characters, Britton has developed a line of mounted alien heads cast in resin, creating a whole sci-fi backstory about the hunted aliens and those who hunt them

for sport (think: DIY *Predator*). Besides playing off the idea of intergalactic trophy hunters, there's a practical purpose to the design of the line. "I heard many people say they were running out of shelf space for their model collections, so I figured, how about making heads that hang on the wall?" For his designs, Britton riffs off of creature forms found here on Earth, even incorporating organic materials, such as animal bones, into his original sculptures. His latest series is of the alien hunters themselves. Dubbed "Boneheads," the first bust, Admiral Enob, is built around part of a bird skeleton, a fish jaw, and a crab shell. While this is a sophisticated sculpt, it's also a good example of how found objects can be

"A model enthusiast can buy kits of characters from low-budget horror flicks and underground comic books, and spaceships from just about every cult sci-fi classic."

assembled and amended with sculpting materials to create impressive-looking models, even if you don't have Britton's sculpting chops.

The Shiflett Brothers (shiflettbrothers.com), Brandon and Jarrod, are another example of kit makers who are also involved in the entertainment business. These Texas brothers got their start doing sculpture work for the video game series *Oddworld*. They've also just teamed up to work with Nimba Creations, a U.K. F/X company that, among other things, worked on Peter Jackson's *King Kong*. To satisfy their kit-creator jones, the brothers have produced a line of resin kits, including the very Tank Girl-esque model Chloe: Aviator for Hire. While many sculptors in the hobby do everything: sculpting, molding, casting, and sales, the Shifletts (and others) do only the sculpting and then sub out the molding and casting to other hobbyists or small-scale commercial casting operations. "Casting is an art in itself," says Brandon, "and for us, finding the right person to mold and cast our work is extremely important."

Most people involved in the garage kit community want to give something back, often in an effort to

Maker

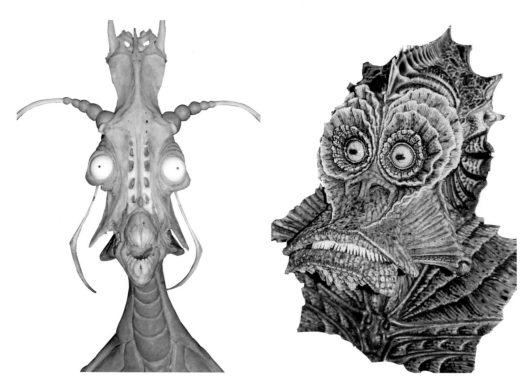

Inspired by books about Bigfoot and the Loch Ness Monster, Dave Britton began creating models at age 15, eventually moving to Hollywood to work as a make-up artist and lab technician. Most of his models are "full size" — the Aqualarian head (right) is 23" tall.

get new hobbyists involved and to form alliances with fellow kit makers. The Shiflett Brothers do this by maintaining "Sculpting with the Shiflett Brothers," the largest sculpting online forum of its type. "Our forum attracts pros and amateurs alike," says Brandon. "Everyone from 13-year-old newbies to pro sculptors working on the *King Kong* film."

Other modelers like Dan Perez include workshops on their websites that teach sculpting, molding, and casting. And then there's the collective policing of recasters. While most garage kit creators don't bother with licensing the subject matter of their kits, they don't take kindly to those who buy one of their kits, create a second-gen mold, and then cast knock-off kits at a third of the price (called "recasts").

"We're a close-knit community," says Terry Webb. "We look out for each other, and when recasters rip one of us off, it impacts us all." He adds, "This is an art form, and those of us involved in it take great pride in what we do."

Gareth Branwyn writes about the intersection of technology and culture and is "cyborg-in-chief" of Streettech.com.

Resources:

Dan Perez Studios Modeling Workshop
danperezstudios.com/workshop.htm
Perez's workshops take you step-by-step through beginning sculpting, mold making, casting, and other skills used in garage kit building. Highly recommended.

The Clubhouse
theclubhouse1.net/
Popular online watering hole for kit creators and builders.

Amazing Figure Modeler
amazingmodeler.com, $28/4 issues
Premier magazine for figure modeling, with how-tos on everything from sculpting your own models to assembling, painting, and basing garage model kits.

➕ For more resources go to makezine.com/05/monsters.

MAKING YOUR OWN MODELS

While garage kit making can be expensive and require serious dedication to the art of sculpting, beginners can get their hands dirty with minimal effort. The easiest way to get started is to buy a casting kit, such as Bare-Metal's (bare-metal.com) Mold Making and Resin Casting Starter Kit. It comes with all of the materials you need to mold and cast a few small models.

Here are the basic steps to model making:

1. Sculpting/Master Modeling
This part of the process obviously requires some talent, but beginners can use "kit bashing" (cannibalizing existing models) to cobble together designs of their own mutant creations. By combining existing kits with plastic stock (found at hobby stores and e-tailers like plastruct.com), you can create a unique model that can then be molded and cast. Figure modeling can also benefit from bashing existing kits as well as using armatures (wire forms that can be bent into desired poses). Sculpting materials vary widely, from Super Sculpey clay, to two-part epoxy putties, to Casteline modeling wax. Every sculptor has a favorite, and different materials are used for different models.

2. Mold Making
Once your master model is complete, you need to create a mold from which the successive kits will be cast. There are a number of methods and materials for doing this, but the most common approach is to create a two-part mold around your model, one half at a time (*see MAKE, Volume 03, page 135, "Squeezable Nightlight"*).

The bottom half of the model is surrounded by a soft, non-sulfur-based modeling clay bed, leaving the top half of the model to be molded in rubber. Next, RTV (room temperature vulcanizing) silicone rubber is mixed and poured over the top to create the first mold. The two mold halves are later joined up securely, thanks to "keys," a series of dimples that are pressed into the clay bed around the model before pouring in the rubber. Air vents and a pour spout are

You will not be able to make anything this cool on your first model-making attempt. Try for a chubby penguin instead.

created by pressing brass rods into the modeling clay on top of the model. A foam-core molding box is built to surround the model and contain the rubber molding material.

Once the rubber has been poured into the box to create the top mold, and is allowed to cure, the mold is flipped over, the clay bed is removed, and the first mold is treated with a releasing compound, which can be as simple as brushed-on petroleum jelly. This prevents the two mold halves from sticking together. Once both mold halves have been created in silicone rubber, casting of your model can begin.

3. Casting
Before casting, the mold halves are treated with a spray-on releasing agent (such as Parafilm mold release). With the two-part mold held securely together with rubber bands, the resin is poured into the spout until the mold is filled up. The most common casting material is a two-part polyurethane resin, such as Vagabond Corporation's Model-Cast (vagabondcorp.com).

Once the casting material has cured, the mold is carefully removed and the cast model is cleaned up, ready for painting.

A crowd gathers as jostling robots struggle their way through Peter O'Kennedy's infinite loop, titled *Escape*.

Re-Booting Art

A new art venue for the robotic age.

By Douglas Repetto

WHAT IS A ROBOT? WHAT IS ART? Those aren't easy questions to answer. Now try defining *robotic art*, or decide what exactly an art-making robot is.

In 2001, I was learning the difficulties of trying to show robotic work in traditional art venues. (It took decades for galleries and museums to embrace video art; robotic and kinetic works, with their sometimes-formidable technical and logistical complexities, have quite a way to go.) But at the same time, robots were all over the popular media. Most of the attention was focused on violent applications, like BattleBots and Patriot Missiles, yet I knew of many artists working with robotics whose strange, subtle, or whimsical ideas had little in common with the remote-controlled chainsaws on TV.

With those two issues in mind, I started a robot talent show email list. Over the next several months, many people contributed ideas for how to create such a show. As those discussions quickly made evident, robotic art is hardly a well-defined genre, and it's not clear what sorts of works qualify. Is an inkjet printer an art-making robot? Is a purely mechanical sculpture robotic art? Our response to these kinds of questions was to dodge them entirely; our open call for entries says: "If you think it's a robot and you think it's art, then send it in." That strategy of inclusiveness has paid off, and some of my favorite pieces have been ones that most liberally interpret both "art" and "robot."

Photography by Douglas Repetto

ArtBots: The Robot Talent Show is now an annual art exhibition (see artbots.org) where an international group of artists gather to share their work and their trade techniques, and give a general audience a chance to see what kinds of answers (and new questions!) are being cooked up in labs, garages, and studios around the world. Each year, our location changes and I invite a couple of new people to co-curate the show with me. (Philip Galanter, an artist and teacher, joined me in co-curating and producing the first two iterations of the show.)

We hope that by tweaking the specifics, we can keep the show open and accessible to a diverse range of people, works, and ideas. There have been robots that draw, robots that dance, and everything from a self-preservation machine to a robot race in an infinite loop. ArtBots is low-budget and hands-on (all of the artists attend and help mount the show), and is dedicated to the idea that technology is ours to use and create, and that making art and asking questions is an essential use of any technology.

Douglas Irving Repetto is an artist and teacher involved in a number of art/community groups including dorkbot, ArtBots, organism, and music-dsp.

1 Wildflower Meadow Glacier **by James Powderly, Michelle Kempner, Tom Kennedy, Todd Polenberg, Brendan Fitzgerald, and Paul Bartlett. Scale model of an enormous artificial glacier moving through Central Park, leaving a path of wildflowers in its wake. The density of flower planting is determined by continuous monitoring of the CO_2 levels in the local environment.**

2 Robots Like H_2O: Photosynthesis Perpetual Motion Machine **by Futurefarmers: Amy Franceschini and Michael Swaine.**

Calculate This!

By Robert Luhn

In the Pre-Digital Epoch, people building bridges, launching rockets, or trying to fill an inside straight used manual calculators. It might've been a snazzy slide rule, or a paper wheel with dials that (when properly aligned) computed a sales commission, the proper valve fitting, or how big a hole a 20-megaton H-bomb would make.

As one observer put it, manual calculators like the slide rule and its kin "radically increased our capacity to perform complex mathematical computations. They literally enabled us to develop our modern world." Here are six classic calculators, ranging from the mind-blowing to the arcane.

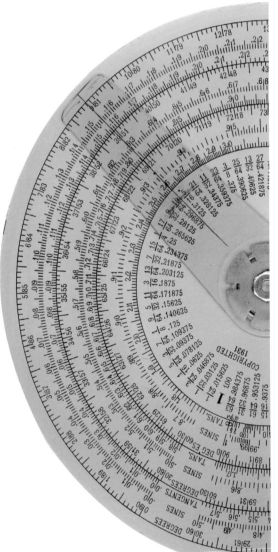

Gilson Binary Slide Rule | 1940

When people think of slide rules, they think of the classic rectangular slipstick. But many engineers, scientists, and merchants turned to this Frisbee-sized binary slide rule from Gilson (a whopping $11 with case). This was the HP calculator of its era, sporting nearly two dozen scales, from trig functions to logarithms to multiplication. But this calculator was rooted in the real world — for merchants, the Gilson could quickly compute retail pricing and add and subtract fractions. For engineers, the Gilson could compute an answer to five figures — not bad for a pre-Pentium device. As with all slide rules, your ability to estimate numbers and keep track of decimal points was key to getting the right answer. (For more calculating, go to the Slide Rule Universe at sphere.bc.ca/test/sruniverse.html.)

Characteristics Calculator for Large and Miniature Filament Lamps | 1952

How long before that incandescent bulb blows? This handy General Electric calculator — probably created for GE salespeople pitching the company's wares — has the answer. Turn the bulb-cum-pointer until it alights on the socket voltage and bulb voltage, and you can quickly learn how much energy the bulb consumes, the relative cost of lighting your hovel with this particular bulb, and how long the bulb will last before it goes nova. (To shed further light on this topic, seek out the Hall of Electrical History at schenectadymuseum.org.)

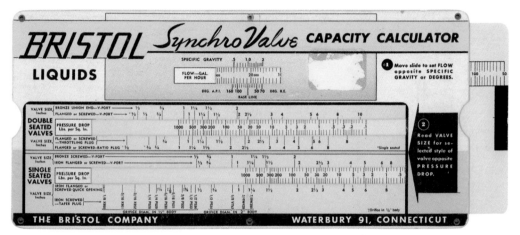

Bristol Liquids SynchroValve Capacity Calculator | 1943

No, this isn't for tuning up your Flux Capacitor. Bristol manufactured sensors, recorders, and other instruments for companies working with liquids and gases, from water to steam to pancake batter. This handy multicolored slide let engineers quickly figure out the right valve type and size for a given fluidic situation. Bristol closed its doors years ago; according to the EPA, the company site is a heady stew of lubricants, solvents, PCBs, and other toxic byproducts. Calculate the cost of *that*.

Poker Pal | 1955

When the chips are down — and you wonder if you'll ever fill that inside straight — reach for Perrygraf's Poker Pal. With a quick yank of the slide, you can determine what hand you need to win (in draw or stud) — and the odds of getting it. You can "what-if" with ease by setting the number of players, number of cards wanted, what's showing on the table, and so on. In business since 1934, Perrygraf is still "THE source for innovative Slide-Charts, Wheel-Charts, Pop-Ups and other dimensional marketing products." (To see its modern-day equivalent, slide over to poker-pal.com.)

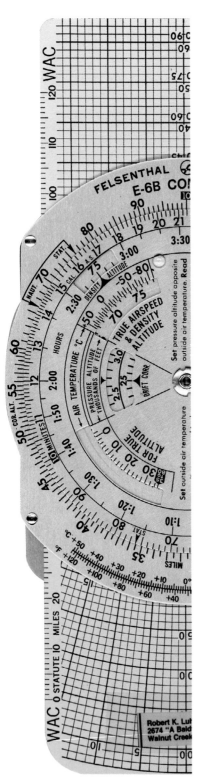

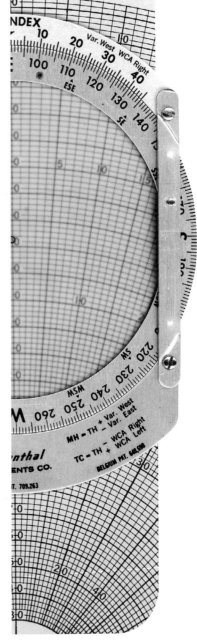

E-6B Flight Computer | 1966

Time, speed, distance, fuel. These are the mantras of the pilot. In the old days (pre-1990s), pilots relied on this device to solve such in-flight quandaries as "If I fly 40 minutes and consume 75 gallons, when will I run out of gas and plow into that mountain?" One side of the E-6B is devoted to such issues, and the other, to figuring out your true direction and speed while correcting for the nefarious effects of wind. These days, you can pick up a digital version of the E-6B for $60 that does a heck of a lot more — but nothing says "veteran aviator" like this baby tucked under your arm. (To see how modern fliers plot their flights, check out the DUATS system at duats.com.)

Today, when everything is programmable, it's reassuring to discover dedicated calculators that are as straightforward to use as a hammer.

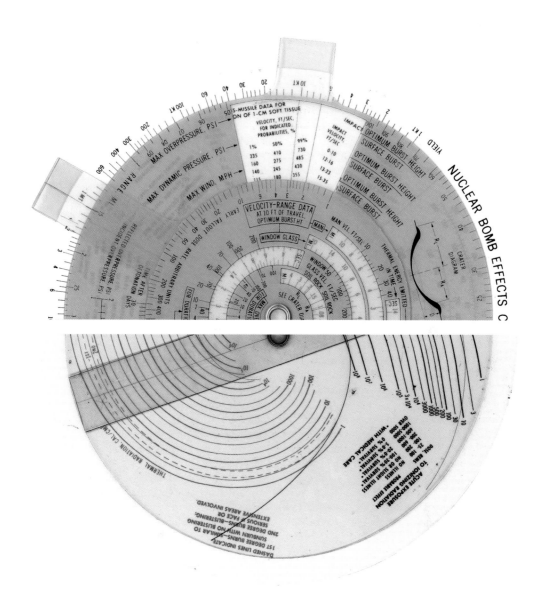

Nuclear Bomb Effects Computer | 1962

What every nuclear family needed in the 1960s. In an era when building a bomb shelter was more fashionable than building a swimming pool, the Atomic Energy Commission contracted with the Lovelace Foundation to create this handy computer. Know the megatonnage of that incoming Soviet ICBM and where it hit? Turn the wheels to those values and you can learn the maximum radius of the fireball, the velocity of burst window glass, your chances of surviving 5,000 rems, and more. Fun for the whole family! (For more history and the scoop on building your own bomb effects computer, skulk over to fourmilab. ch/bombcalc).

Robert Luhn is an executive editor at O'Reilly Media. He collects antique machines and small, unattractive animals.

Saul Griffith

PRE LOVED

OR, WHY USED CAN BE SO MUCH BETTER THAN NEW.

MY GIRLFRIEND RECENTLY SAID, "LET'S get a dirt bike."

"Woohoo!" I said. Then, "Wait, why do you want a dirt bike?"

The motives, according to her, were a string of recent events. 1. *Deep Impact*, the movie: remember the scene where the two kids escape the tsunami caused by the meteor impact with the Earth? 2. The newspaper coverage of traffic jams of people trying to get out of the way of Hurricanes Katrina and Rita. 3. We live in the East Bay of San Francisco in a low-lying area only a few miles from a bunch of major fault lines, with the potential for still more post-disaster traffic jams.

Of course, it may have been none of these reasons. She is an aesthete, after all, and her choice to Netflix the fabulous film *On Any Sunday* was probably the major influence, however subconscious. It's a 1971 documentary on dirt bike racing bankrolled by, and starring, Steve McQueen, produced by the same filmmaker who made the cult surf classic *Endless Summer*. We both loved the movie, particularly the simple beauty of dirt biking before advertising and logos made every amateur look like a lime green advertisement for extreme sports drinks.

One of the most beautiful aspects of *On Any Sunday* is the footage of trials-riding legend Malcolm Smith participating in an amazing motorcycling marathon. A multiday event where the rider was expected to ride the same bike, all components included, across a long and tortuous route including the mountains and roads and mud and dirt and deserts of Spain —

sort of a dirt-bike Tour de France. The only person allowed to fix the bike during the ride is the rider himself, right down to changing his own tire. As the ride is a time trial, one must complete each section within a given time period, including repairs, so it pays equally to be an awesome rider and a talented and intuitive repairman. What a refreshing format: a competition for renaissance men, people who understood their machinery as much as having the testosterone to ride it fastest. I would love to have seen the fencing-wire and duct-tape solutions that got the winning machines over the finish line. I say winning machines because the race was not for first place, since anyone who finished with original bike within the time limit got a medal.

I could see why my girlfriend might have been seduced, but I was still suspicious. I broached the subject with one of my dear friends. "A dirt bike?" he said. "Your girlfriend is suggesting you buy a dirt bike?! Don't worry about the motives, just say yes, yes, yes!"

I started looking at dirt bikes online. I had to filter my search by cubic engine capacity as much as by space for two riders and the aesthetic requirements of my lady, because after all, if you are going to escape the apocalypse, you must escape it in style. I wanted something I could fix with minimal tools and equipment — this ruled out an awful lot of the newer bikes, which were too expensive anyway, and many of the older bikes, which were terribly troublesome looking. Eventually, we found the Bultaco, which, it so happens, is a Spanish machine whose heyday was in the 1970s. I quickly found one nearby on Craig's List (craigslist.org). On the web, I also found a plethora of sites about passionate Bultaco hobbyists, a good sign I'd find community support and hints on keeping the machine alive. Although the last manual was printed 20 years ago, I'd still be able to train myself in the finer arts of Bultaco-owning.

Buying anything secondhand, particularly a 30-year-old machine *designed* to be abused, is always risky. I fully expected to be buying it from a 30-year-old hooligan who'd thrashed and beaten it and was now disposing of it. I was delighted, therefore, upon approaching the Cupertino driveway to see

The Bultaco's glory days were also the heyday of engineering renaissance-men.

Photograph by David Sobo

a middle-aged man mounted atop this machine with the confidence and man-machine respect of Malcom Smith; he was having a little "goodbye ride," likely reminiscing, before we arrived. I was even more delighted when I entered his workshop. It was the epitome of the phrase "a place for every tool, and every tool in its place." This was a bike that had been nurtured and nourished and loved by an artisan.

Curiously, an old ambulance sat in the driveway alongside a new Mustang and vintage kit car. While I discussed the fine points of oil changing and the reserve fuel tank of my girlfriend's new Bultaco with Tom, she found out from Tom's wife that he was a keen competitive amateur motorcyclist, and that he had converted the ambulance so it was both bike transport and mobile workshop. Legend has it he was known as the only rider to *arrive* at races in an ambulance.

Tom's expertise with bikes extended to less tricked-out transports as well: I was happy to learn the two-tie-down trick for loading a dirt bike in the back of a pickup (one to each side of the handlebar, compressing the front suspension and locking down the front wheel — the rear then doesn't need tying). While adding this arcane detail to my repertoire, I asked knowingly, "You're an engineer?" To which he acknowledged, "Yup." Sometimes a word has

curiously high bandwidth; this one signaled that he knew I too was an engineer and that I'd care for his machine as he had, not polishing it to a shine, but rather developing a symbiosis, understanding the rattles and what they mean before they develop into more terminal mechanical problems.

So now I have another old machine in my care. It will require periodic tuning (and yay for me, periodic

"This was a bike that had been nurtured and nourished and loved by an artisan."

"test" riding) to make sure it is ready should the big wave or crack in the ground ever come. I choose to believe the disasters are unlikely and that now I'll get to spend the odd "any Sunday" tuning the carburetor and riding around the block imagining myself in Spain in a man-machine marathon.

Saul Griffith thinks about open source hardware while working with the power-nerds at Squid Labs (squid-labs.com).

JACKHAMMER HEADPHONES

By Tim Anderson

AND A DISCOURSE ON BLIND MEN, CHIPMUNKS, WHALES, AND THE FUTURE.

I made my first pair of jackhammer headphones about 15 years ago while doing volunteer work for the blind. Many organizations are devoted to helping the blind, including the Library of Congress (LOC) and the Minnesota Braille Press, where I was a volunteer.

The LOC mailed blind people audio books on half-speed cassettes that played on special orange tape recorders. I used to repair these machines along with a group of retired telephone company workers. Other volunteers, mostly actors, read books onto tape in many sound booths. We volunteers were allowed to check out audio books ourselves from the huge collection. The first one I got was *Moby Dick*. Rather than using one of the clunky LOC machines, I used my (no-name brand) walkman. I reduced my walkman's motor speed until it seemed about right and the narrator sounded like a chipmunk. Even so, it took a long time for Mister Chipmunk to recite the book on about 30 cassettes. (In my memory, the book is like this: "Oh boy! Here's a bunch more facts about whales! Blah blah blah blah!")

At that time, I was driving a VW bus with no muffler and hardly any body left. The body had rusted away from Minnesota road salt. On the highway, it was so loud in the cab that I wore jackhammer earmuffs. I wanted to hear what happened to Captain

Ahab despite the noise, so I shoved headphone speakers into my ear protectors, making the first pair of the headphones that are the subject of this article.

FUNCTION IN A NOISY WORLD

This project is the easiest and probably most useful thing I've ever designed. I'd have trouble functioning in this noisy world without them. These homemade hi-fi headphones work as well or better than Bose or Sony noise-canceling headphones, according to friends who have tried both. Unlike the commercial models, these block outside noise instead of canceling it. They save your hearing by blocking outside noise and allowing you to play your music at a lower volume.

I've been making these for many years now and using them daily. People sometimes ask, "Isn't it dangerous not being able to hear?" Maybe. Ask a deaf person. I've never had a problem. Lots of my friends use them and no harm has come to anyone.

When driving cross-country, I check out 5 gallons of audio books from the local library to listen to on the road. It's the only thing that works to keep me awake and (somewhat) sane on a long drive. When I get to my destination, I mail the tapes back to my home library.

MAKE YOUR OWN HEADPHONES

Cost: $20 **Time to Make:** One Minute **Complexity:** None

The Three Ingredients

1) **Industrial ear-protection earmuffs from McMaster-Carr (mcmaster.com) or another vendor. These are Peltor model H10A, the best I've found. They are the quietest earmuffs available (30dB) and have a lip inside that holds the speaker in place.**
2) **Any airline or walkman headphones of the one-wire-per-ear variety.**
3) **A cutting tool.**

Step 1: Cut off the head loop.

Clip, cut, or chew off the plastic loop that connects the headphone earpieces. Note the right-angle-style phono jack. This type is good. It is less likely to damage your walkman than the straight kind. The straight ones tend to pry the audio-out socket loose from its internal solder connections. A limp cable is also desirable.

Step 2: Shove a speaker into an earpiece.

Let me say this again, Peltor model H10A earmuffs are perfect for this. They have a rim inside that holds the speaker in place.

Finished!

Repeat with the other speaker, and you're done! Enjoy!

Tim Anderson, founder of Z Corp., has a home at mit.edu/robot.

Photograph by Tim Anderson

Tips and Tricks

Helmet-Mounted Headphones
You can see my bike and kitesurfing helmet at www.makezine.com/05/heirloom. In the photos, I'm trying it out with Peltor helmet-mounted headphones. These hangers are slightly more comfortable than wearing the headband under the helmet, but I like being able to take the helmet off and keep the headphones on. So I went back to using the headband-style headphones under the helmet.

The "Double Diaper" Trick
In a really noisy environment, use foam earplugs in addition to these headphones and turn up the volume to a comfortable level. Good foam earplugs can block 27dB. Worn correctly, these headphones block about 30dB of outside noise. In a perfect world, you'd put them together and block 57dB, but I'll leave the numbers to the scientists. It's impressively quiet.

What to Listen to?
My preference is audio books. After listening to 500 or so books on tape, I find myself with quite a store of useful knowledge. Most public libraries have a good selection of audio books. They'll also purchase items you request if they don't already have them.

TIPS AND TRICKS
FROM MAKER TO MAKER

Who doesn't appreciate a really good tip now and then? Especially the kind, as one reader put it, "that changes your life." Whether it's something as practical as revealing a way to remove scratches from glass, or something more creative, like how to make an origami envelope, we all rely on our friends and neighbors to tip us off to the new and the good. —*Arwen O'Reilly*

Tip Your Glass

Tom Bridge saves you from embarrassment: "When you're knocking back a pint at the local bar, you look like an absolute idiot when that little cocktail napkin is stuck to the bottom of your glass. Grab the salt shaker from the guy next to you and dash out a bit on your napkin. It will soak up beer and condensation, but not stick to your glass. Now you can drink a beer without looking like an unwashed cretin." (That's why some of us drink at home.)

Inner Tube Tie-Downs

What won't MAKE columnist **Saul Griffith** ever be found without? Far more useful than a bungee cord, he uses bicycle inner tubes for everything, from strapping broken parts together when something fails away from the workshop, to lashing kitesurfing gear to the back of his bike, to fixing the fan belt of a broken-down Chinatown bus. Most bicycle repair shops throw out used tubes and will be happy to pass them along to you.

No More Scratches!

Scratched CRT monitor? Can't see through your glasses? Rather than replace the glass or buy an expensive scratch-removal product, just use a little bit of toothpaste (make sure it's paste, not gel, and one with baking powder is best). Rub over the scratch with some lint-free cloth "in a small circular motion with moderate pressure for a minute or two" and wipe off, says MAKE reader **Paul Short**.

Better Shoelace Tying

James Arlen wants to make sure you never trip over your shoelaces again. Tie your shoes the normal way, but after you make the first loop, wrap twice around the first loop before pushing through the second loop. "Yes, it's that simple," he says. "The knot won't come undone, there is no need to 'double knot,' and you can still undo the laces just by pulling the free ends. This works with bike shoes (notorious for coming undone), running shoes, dress shoes, boots, and especially little kids' shoes!"

HANDY WEBSITES

DIY Laptop Feet

If your poor laptop sheds feet like cherry blossoms in a light spring rain, here's a thought for you. This page has simple DIY instructions for making even sturdier replacement feet at less than one-tenth the price of commercially available ones. All you need are a package of self-adhesive surface protectors (the kind you put under lamps and such) and a sturdy hole punch.

the-wabe.com/notebook/laptop-feet/

Tricks of the Trade

A whole website of tips offers "professional secrets from those in the know." Each blog entry takes the form of a reader-submitted trick of the trade. It has advice by everyone from window washers (wash each side in a different direction so you know which side streaks are on) to soldiers in Iraq, who are using Silly String to check for bomb tripwires (spray in a dark room and it clings to even the tiniest of wires). Practical and beautiful, the site is yet another example of why we are greater than the sum of our parts. tradetricks.org

The Shelf Life Index

Ever wonder just how long your mayonnaise will really keep? How about that old can of soda? What's the difference between something just tasting bad and actually making you sick? *Real Simple* has a great list of 77 common foods and their shelf lives to solve your last-minute panics and late-night musings.

makezine.com/go/shelflife

Give Away Unwanted Stuff

Freecycle is an international network of local groups who use email listservs to post about things they want to give away — used couches, exercise equipment, broken-but-fixable electronics, film canisters, cardboard boxes, plant cuttings. Everyone wins here: givers get more garage space, takers get something for free, and landfills are conserved. Plus, you get to meet your neighbors.

freecycle.org

GPS Stamping for Photos

The Motorola i860 on the Nextel network works with an application that allows you to upload geo-stamped camera phone pictures to a website. "There's even an Easter egg that lets you upload directly to flickr.com," says Allen Smith. "The phone adds tags of the city, state, and zip code where the picture was taken, as well as a Mapquest link. To enter your Flickr email, press # *your pin* 8 7 5 0 # from the first start-up screen, and a pop-up screen lets you enter your Flickr email and password."

geosnapper.com/mobile.php

Have a tip for MAKE readers? Send to tips@makezine.com.

Origami Envelopes

Letter writing may be quickly becoming a lost art, but **Mark Brown** has found just the way to spice it up again. Flying Pig, a website of whimsical paper animation kits, also has a free downloadable PDF of instructions for making a paper envelope. "I've seen a lot of different ways to fold an origami envelope," says Brown, "but the simplest and most elegant, bar none, is this one. Too cool."

flying-pig.com/pagesv/envelope.html

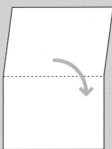

Start with a piece of 8.5x11 or A4 paper. Fold the paper in half horizontally and make a crisp crease. This is your center line. Open up the paper again.

Fold over the two diagonals and carefully line up the edges with the center line.

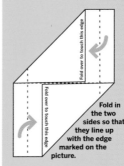

Fold over to touch this edge

Fold over to touch this edge

Fold in the two sides so that they line up with the edge marked on the picture.

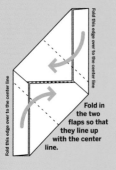

Fold this edge over to the center line

Fold this edge over to the center line

Fold in the two flaps so that they line up with the center line.

Tuck under

Tuck under

Tuck under

Tuck the ends of the two flaps into the two small pockets. That's it. Turn your envelope over and address it.

Flying Pig
Broughton Moor
Maryport
Cumbria
CA15 7RU

Be recursive! Print out these instructions, make them into an envelope, and mail them to a friend.

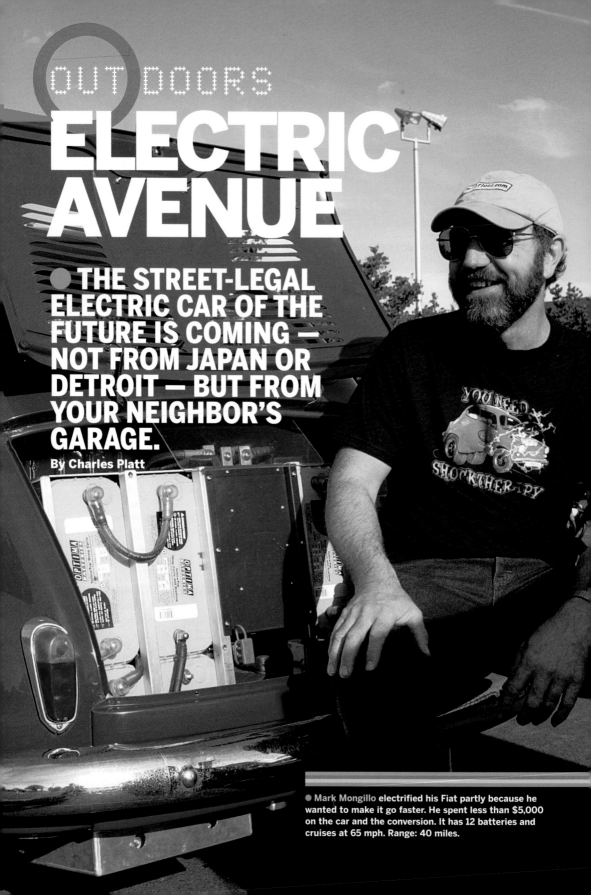

ELECTRIC AVENUE

● **THE STREET-LEGAL ELECTRIC CAR OF THE FUTURE IS COMING — NOT FROM JAPAN OR DETROIT — BUT FROM YOUR NEIGHBOR'S GARAGE.**

By Charles Platt

● **Mark Mongillo** electrified his Fiat partly because he wanted to make it go faster. He spent less than $5,000 on the car and the conversion. It has 12 batteries and cruises at 65 mph. Range: 40 miles.

ON A FRIDAY NIGHT AT THE
Portland International Raceway, teen-age greasers in tricked-out muscle cars gather to strut their stuff. Out in the staging area, purple smoke drifts from burnouts under mercury-vapor spotlights while loudspeaker announcements echo above the rumble of blown V8s. Here in this all-American ritual fueled by gasoline and testosterone, I'm sitting in the passenger seat of a totally incongruous interloper — a cube-shaped postal delivery van powered entirely by batteries. The owner, Roderick "Wild Man" Wilde, did the conversion for a one-hour show on the Discovery Channel. His motive was simple: to prove that electric vehicles don't have to perform like golf carts.

As I perch in the passenger seat beside Wilde, I am uncomfortably aware that if he presses the pedal to the metal, the trio of electric motors driving all four foot-wide dragster radials will be juiced with 2,000 amps at 240 volts from 40 Exide XCD batteries. That's almost half-a-megawatt, enough electricity to sustain 50 average homes.

"When I was a kid," Wilde remarks, "my father was a service manager for a Chrysler dealership, and he told me that gasoline motors are stupid. The piston has to go up, stop, and come back down — it's very inefficient." He smiles reflectively. "If I can take this postal van, weighing more than two tons, and hurl it down the quarter-mile, maybe people will revise some of their anti-electric prejudices."

A few minutes later, after I step out of the van, he peels out with tires smoking and accelerates to 80 miles per hour. He's disappointed; he hoped to break 100.

In an era where alarmists speak of "peak oil" and gas prices have jumped 50% within a year, EV advocates see an unprecedented opportunity to deliver a message that has been persistently misrepresented or misunderstood. Electric vehicles do not have to be overpriced and overweight, like GM's EV-1 or Toyota's RAV4. Nor do they have to be flimsy absurdities such as the Gizmo or the Twike, which look like refugees from reruns of *The Jetsons*.

Modern batteries and high-amperage controllers can satisfy any reasonable need for speed — in a car that looks like a normal car. In fact, if you have minimal mechanical skills, you can do your own gasoline-to-electric makeover for a mere $7,000, acquiring an automobile that has zero emissions yet still boasts enough low-end torque to burn rubber or (more sensibly) merge with traffic doing 75 on the freeway.

> **" If you have minimal mechanical skills, you can do your own gasline-to-electric makeover for a mere $7,000, acquiring an automobile that has zero emissions yet still boasts enough torque to burn rubber. "**

This seemingly impossible dream of a high-performance ride within the constraints of ecotopian frugality raises an obvious question: if EVs can be so easy to build, cheap to run, and fun to own, why aren't we all driving them? Even more perplexing, why have today's auto manufacturers abandoned EVs in favor of hybrids, which deliver less performance with greater complexity at a higher price, while still emitting greenhouse gases from their tailpipes?

ON THE SATURDAY MORNING
after Wilde's racetrack adventure, he tows his mail truck into downtown Portland, where a miscellaneous assortment of vehicles is gathering at the parking lot outside a Village Inn diner. This annual show of street-legal electrics is being sponsored informally by local resident John Wayland, whose innovative backyard engineering and persistent proselytizing have earned him guru status in the EV world. Wayland is displaying three of his own creations: a commuter-friendly renovated 1972 Datsun, which looks sufficiently immaculate to be a show car; another 1972 Datsun dubbed the "White Zombie," which is the world's quickest street-legal electric car, running a quarter-mile in just over 12 seconds; and — as if the man needed to prove that he has a sense of humor — an electrified lawn tractor enhanced with a thumping stereo system.

Wayland is cheerful, amiable, and relentless, even when a skeptical elderly visitor utters the inevitable complaint: "Do you have to plug these things in?"

Well, duh, yes, you have to recharge an EV, just as you would recharge a laptop, an iPod, or a cellphone. And so we get to the inevitable follow-up: "How far will it take you?"

Almost all the vehicles being shown here share the same limit: around 40 miles. Inevitably, this rouses doubt and dismay, since a tank of unleaded typically lasts for 200 miles or more. The fact is, though, most of us seldom use this capacity. Half the homes in the United States contain two or more cars, and the number-two car performs local errands such as picking up the kids from school, commuting to work, or visiting the supermarket. For this kind of duty cycle, an EV with a 40-mile range is ideal.

When big companies such as Ford and GM offered EVs, they didn't even try to pursue the short-range paradigm. Instead, they attempted to guarantee as many miles as possible by loading their vehicles with massive battery stacks that devoured interior space and crippled performance. The result was an unhappy compromise that pleased no one.

Mindful of this foolishness, backyard builders acknowledge the limits of today's battery technology and sacrifice some range in order to save weight, mimimize expense, and maximize drivability. Here's a classic example: a cute little green vintage Fiat, the size of an old VW Beetle. "I've always had a thing about Fiats," comments its owner, Mark Mongillo, "but they weren't fast enough as gasoline cars." After cramming a couple of battery stacks into the rear engine compartment, he has a livelier vehicle that still seats four adults. "I bought it for about $2,500," he says, "and put about $2,000 into the conversion, with a lot of used parts bought over the internet and through friends. It took me about a year of tinkering — a few hours here and there. I mostly did it because it's fun."

Mongillo drives me around some back streets, and I notice that the motor is so tolerant of speed variations that he doesn't have to touch the gearshift. "Second gear is good for up to about 45 miles per hour, and third gear takes me up to 75," he explains. Nor does he need to slip the clutch when starting from a red light, because an electric doesn't have to rev up in preparation for taking off.

Back in the parking lot, I inspect Rick Barnes' 1986 Chevy Sprint, the dowdiest vehicle on display. "I was going to convert a Yugo," he says, "but I found this already converted, for sale in Seattle." He says there is an active secondhand market for EVs, because no one wants to throw them away. He didn't bother to pimp his ride for this car show,

> **❝Cruise past gas stations with the same feeling of superiority as an ex-smoker bypassing a display of cigarettes in a convenience store. 'Middle Eastern oil? Sure, I used to burn it — but I quit.'❞**

because the Sprint is not a show car. He commutes in it every day to his job at a local Intel chip factory.

At the other extreme, Russian-born electronics engineer Victor Tikhonov displays an ongoing research project: a 1991 Honda CRX powered with state-of-the-art lithium-ion (Li-Ion) batteries, which offer greater range but have a higher internal resistance that limits amperage and acceleration. To compensate, Tikhonov has added a massive array of capacitors, visible beneath the Honda's fastback rear window. "There are 160 of them at 2,700 farads each," he tells me.

Another lithium-ion advocate is Jukka Jarvinen, who flew here from his native Finland. Jarvinen is tall, thin, intense, and dogmatic in his passion for all things electric. He built a lithium-ion-powered motorcycle, and recently obtained government funding for his own EV battery development company. "You get about 25,000 miles from one lead-acid battery pack before it must be replaced, but lithium-ions easily give you 150,000 miles," he claims. "That's about 10 years from one set of batteries, with 160 miles per charge." He believes that lithium-ion prices will plummet, since the Chinese have invested $2 billion in battery

1. Roderick Wilde's 500,000-watt postal van arrives on its trailer to compete with street-legal dragsters at the Portland International Raceway.

2. Tom True (left) enlisted assistance from EV veteran Don Crabtree (right) when he wanted to electrify his Datsun Z. While all the original accessories still work, including power windows and radio, the gasoline engine has been replaced with a cluster of three electric motors.

3. On the concrete floor of John Wayland's garage are the components for a typical EV conversion: eight 12-volt lead-acid batteries, a forklift truck motor and controller, recharging module, DC-DC converter, mounting hardware, and drivetrain coupler. Total cost including batteries: $7,000.

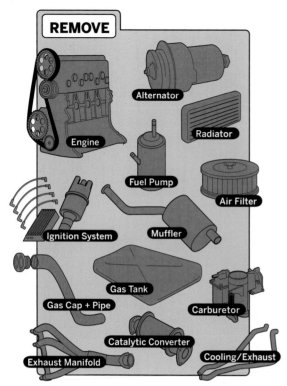

REMOVE

Alternator

Engine

Radiator

Fuel Pump

Air Filter

Ignition System

Muffler

Gas Tank

Gas Cap + Pipe

Carburetor

Catalytic Converter

Exhaust Manifold

Cooling/Exhaust

INSTALL

Batteries
Eight 12-volt lead-acid
batteries. $500-$2,000.

Motor Controller
Used to moderate
power. $700-$1,500

Motor Mount
Aluminum plate.
$100

Motor
Refurbished part
from a forklift.
$500-$2,000

DC-DC Converter
Changes output
voltage. $150-$300

Charging Module
Recharges batteries.
$1,500-$3,000

Ceramic Heater
1,500W home heating
element. $50

research and have lithium-rich ore deposits in Tibet.

Still, most EV builders prefer the reliable simplicity of lead-acid batteries and refurbished DC forklift truck motors. Pioneers such as Wayland and Wilde figured out 10 years ago that you can safely force 96 volts through a 48-volt motor if you turn the commutator ring to change the timing. Jim Husted, who ran a business refurbishing forklifts for 24 years, was stunned when he discovered what the EVers were doing. He is now a true believer, hanging out with the nerds and spreading the gospel with wide-eyed intensity.

"This is very exciting for me," he says. "I want to get more of these cars out on the road. I'll work within any EV builder's budget to make his car better. Send me a motor and I'll send it back with variable timing. I want to make a retrofit kit, a user-friendly way to do what these guys have been doing on a much more technical basis."

He sounds giddy with the possibilities. "There are thousands of different models of DC motors that have been built in the past 40 years. Some would be ideal for motorcycles. And I'm looking to wire a pair of motors so they can go from series, to series-parallel, to parallel to make a three-speed transmission. I'm proud to be a part of all this."

AFTER THE SHOW, I VISIT THE
two-car garage behind Wayland's suburban home and ask him to show me the parts that are needed for a conversion. It takes him only 10 minutes to gather and array eight 12-volt batteries, a motor (refurbished from an electric forklift truck), a controller to moderate the power, some heavy-duty copper cables, and a recharging unit. With minor additional items such as a 5,000-ohm potentiometer that fits under the gas pedal and costs less than $10, this is all you need. The lesson is obvious: EVs are ideal for backyard builders because they can be really, really simple.

The first step in a conversion is to find a "donor vehicle," usually a lightweight compact gasoline-powered automobile from a scrap yard. You throw away almost all of its mechanical components: the engine, radiator, hoses, carburetor, air cleaner, ignition system, starter motor, exhaust pipe, muffler, catalytic converter, fuel pump, and (of course) the gas tank. An electric motor needs none of these peripheral devices. It is also one-quarter the size of a four-cylinder gasoline engine, about four times as efficient, and far more reliable. An electric vehicle needs no external cooling system, filtering systems, or oil supply.

Illustration by Tim Lillis

● John Wayland's immaculate 1972 "Blue Meanie" Datsun includes protective plastic covers over additional batteries under the hood, alongside a green "Godzilla" controller.

With one massive motor under the hood driving the front wheels, and two more for the rear wheels, the postal van can smoke all four of its foot-wide radials.

After you get someone such as Jim Husted to tweak your motor (or buy a pre-tweaked motor from a source such as EV Parts (evparts.com), you'll need to attach it securely to the frame of the car. This should be simple enough, since you don't need rubber mounts. To attach the motor to your car's transmission, you can get a local metal fabrication shop to create a suitably drilled quarter-inch aluminum plate for maybe $100. You'll also need to get a hub adapter to connect the motor shaft with your existing flywheel and clutch. These are available for many types of cars on various EV parts retailers on the web.

You'll wire your batteries with caution and respect, since their high amperage at 96 volts is hazardous. It's a good idea to invest in some EV conversion how-to books such as *Convert It!* by Michael Brown or *Build Your Own Electric Vehicle* by Bob Brant, or go online to discussion sites where technical knowledge is exchanged freely.

Expect to spend between $5,000 and $10,000, or more if you want extras such as separate motors to power an air-conditioning compressor, power steering, or power brakes. Most EVers choose small, light cars in which power assists are not necessary.

During the winter months, the efficiency of an electric motor actually counts against it, because it generates virtually no waste heat to warm the car interior. Still, you can use the existing car heater

> ❝ **You get about 25,000 miles from one lead-acid battery pack but lithium-ions easily give you 150,000 miles.** ❞

if you insert a ceramic element from a 1,500-watt (minimum) heater of the type sold for home use. When you run it from your DC batteries, it will deduct about 10 to 15 percent of your range.

Cold weather will also affect the batteries themselves. At 32 degrees F, lead-acid electrochemical reactions create only half as much power as at 80 degrees. Still, you can line your battery box with styrofoam, and keep your car in a garage overnight. Alternatively, according to Wayland, "If you can afford a set of NiCads, they have a cycle life in the thousands, and they're on the secondhand market now. At 20 degrees, they still have 98% of their power. Nickel-metal hydrides and lithium-ions also are not significantly affected by cold."

No matter which batteries you use, they'll require some nurturing. You'll try to avoid discharging them completely, and you'll use a properly engineered unit to recharge them. With a small amount of care, they'll last several years, and you'll have the pleasure of driving a truly unique, silent, nonpolluting vehicle. You'll cruise past gas stations with the same feeling of superiority as an ex-smoker bypassing a display of cigarettes in a convenience store. "Middle Eastern oil? Sure, I used to burn it — but I quit."

Of course, electricity has to be generated somewhere, but ideally it may come from hydroelectric installations or wind turbines, and even if it originates from fossil fuels, a large power plant is more efficient and cleaner than a small internal-combustion engine. Also, it may use domestically mined coal, of which we have copious reserves.

Your EV may not have the megawattage to burn rubber on a drag strip in the style of Roderick Wilde's demented postal van, but it will be more spirited than a hybrid while also being simpler, more reliable, cheaper, and more efficient. "The only reason that hybrids are catching on," says Wayland, "is they get people over their fear of not having enough battery power. People can still go to the gas pump. They have gas to rescue them." He shakes his head sadly. "It's a crutch! Hybrids are training wheels for the masses!"

From an EV perspective, hybrids are not only lame but crippled, since their batteries can only be recharged via the car's gasoline engine. Perhaps in an effort to avoid the "EV stigma," manufacturers omitted any option to recharge them from an external source. If this feature were added, Wayland points out, "You could drive a hybrid in the city all week long and never turn on its gasoline engine. But you'd still have the engine when you need greater range."

How long will it take for large auto companies to stop thinking that consumers will be afraid of plugging a car into a wall outlet?

Fortunately, we don't have to ponder this question because the components on Wayland's garage floor offer an easy option. If your daily driving habits fall within a 40-mile limit, you can have your own rationally engineered EV right now. All you have to do is make it yourself.

HYBRID OWNERS PLUG IN

Why don't hybrid cars have charging plugs as well as gas tanks? The auto industry's stated rationale is that the larger batteries needed to store a useful "full tank" of electricity are too heavy and expensive. But many hybrid owners are doing plug-in electric hybrid (PHEV) conversions themselves, and a nonprofit organization, CalCars, has formed to support them and to lobby for PHEV development by automakers.

To convert the popular Toyota Prius, you can replace its 1.3kWh NiMH battery with a 9kWh battery pack, add a power inverter and a plug, and tweak the control system so that the gas engine kicks in at a higher speed (or can be disabled entirely). The result is a pluggable Prius that can make trips around town without using any gas at all, although it won't reach highway speed in all-electric mode.

—Paul Spinrad

RESOURCES

EV discussion and advice:
madkatz.com/ev/evlist.html

Photo album (more than 600 home-built EVs):
austinev.org/evalbum

Parts for EVs:
evparts.com, ev-america.com, manzanitamicro.com, cloudelectric.com

More EV links:
pages.prodigy.net/noela/new_page_2.htm

Victor Tikhonov's "high-end AC drive systems":
www.metricmind.com

North American Electric Drag Racing Association:
www.nedra.com

Charles Platt has been a senior writer for _Wired_ magazine and has written science fiction novels such as _The Silicon Man_.

DIVE, DARN IT, DIVE!

IT'S A MATTER OF SINK AND SWIM AT THE 8TH INTERNATIONAL AUTONOMOUS UNDERWATER VEHICLE COMPETITION.

By Larry Harmon

A Navy diver unhooks Georgia Tech's autonomous robotic submarine for a trial course run at the Space and Naval Warfare Systems Center training pond in San Diego. The team placed 12th in a field of 19.

IT'S 11:30 P.M. ON A FRIDAY NIGHT in San Diego and a group of college guys from a handful of different universities are hanging out in bathrobes and swimsuits by the pool at a resort hotel on the harbor.

But it's not what you think. Instead of chasing girls after beer chasers, these guys are laid out on recliners, writing last-minute programming changes on their laptops, or they're in the pool tinkering with large submarines that look like they were built from erector sets, PVC pipe, and spare PC components.

The students — a mix of 19 teams representing universities and trade schools across the U.S. and Canada, along with a Bay Area high school team and even one from India — are in town for the 8th International Autonomous Underwater Vehicle Competition, held at the Space and Naval Warfare Systems Command (SPAWAR) training pond overlooking the Pacific Ocean.

At stake is $20,000 in prize money, of which $5,500 will be given to the first place team. Considering the cost and time to create a submarine, money isn't the motivating factor. It's all about the bragging rights.

The AUV competition involves three tasks. Once a vehicle is lowered by a small, portable crane into SPAWAR's murky pond, it must submerge by itself and proceed to travel approximately five yards to an underwater gate, which it must enter. This will prove the vehicle is capable of traveling in a straight line.

Next, each robotic submarine must locate a flashing beacon and bump into it to simulate an underwater docking.

Finally, the vehicle must locate a length of zig-zagged pipe laid out on the bottom of the pond and follow it until it finds a large gap. Once it locates the gap, it must drop a marker identifying the missing section. The final segment requires the submarine to listen and locate an acoustic pinger in a marked target zone and surface to be recovered.

At the resort pool, the teams are doing trial runs of the pipeline simulation and tinkering with their machines on the pool deck.

Andy Linnenkohl, an electrical engineering major from Southern Polytechnic State University, located just outside of Atlanta, says his team's chances of winning are about as likely as a high school football team beating the Atlanta Falcons.

His football reference is way out of place among the talk of vector thrust, pressure sensors, and sonar pulses, and it's quickly apparent why he stands out from the other students at the pool.

"I used to be a police officer. We had the big remote-controlled robots for the bomb squad, which was the first time I ever messed with any type of robots before, and it captured me instantly. We had four guys on the bomb squad, all about my age, and when we were alone with that control panel, it looked like a bunch of guys sitting around playing

> **❝ Each vehicle sports plenty of zip ties and duct tape, evidence of quick fixes after going through damage-inducing practice runs in the pool. ❞**

a PlayStation game. We'd be sitting there jerking the joystick back and forth, and the only thing missing was a fire button."

All of the teams are at the pond by 7 a.m. the next morning for the qualifying heats. There are 19 tents forming an L-shape surrounding the pond, organized by last year's placing. The top five tents house the odds-on favorites: MIT, followed by Cornell University, Québec's École de Technologie Supérieure, University of Rhode Island, and Duke University.

At the bottom of the list, mostly because it's their first time competing, are North Carolina State University, Indian Underwater Robotics Society, Georgia Tech's Marine Robotic Society, and DeVry.

Like artists asked to paint the same portrait, but with widely varying skills, experience, and budgets, each of the 19 teams approached the challenge uniquely.

"That's the beauty of working in the water," says Daryl Davidson, executive director of the Association for Unmanned Vehicle Systems International (AUVSI), which organized the competition.

"When you think about the aircraft industry, it's very specific and very elegant," he says. "That's not the case here. Some of them do have very elegant things to them, but it's not so much. When they get into the propulsion systems or battery systems,

Photography by Eric Rife and Paul Hansen

some of them are using very small computers, so they've been able to pack everything down into a small case. One team has a boxy design that looks like Rhode Island. If you have a long, streamlined system, it takes less energy to push it through the water, but once it gets going, it's going to have a harder time adjusting and correcting. Some of these round vehicles can pretty much turn on a dime."

The bigger the budget, the more intricate the vehicle, and it was obvious that the Indian team didn't have much of one.

In fact, team member Anuj Sehgal said they had a total budget of $1,000, which included travel expenses to San Diego. To save money, the entire team was staying at a relative's apartment in the low-rent suburb of El Cajon, approximately 20 miles east of San Diego.

They bought a majority of the vehicle parts after arriving, including the main body, which was a black, squarish, waterproof carrying case complete with handle. Bilge pump motors powered four thrusters — one facing up and one facing down on each side of the case, giving it the appearance of a gyroscopic batwing.

There wasn't enough money in the budget for any type of camera or listening device, which means the vehicle really has no way to complete any of its missions.

No matter what the teams spent, each vehicle sports plenty of zip ties and duct tape, evidence of quick fixes after going through a couple of damage-inducing practice runs in the pond. If a vehicle was having a problem with buoyancy, a handful of washers on a bolt or a plastic Coke bottle filled with hardened insulation foam would balance the craft.

But buoyancy is the least of their problems. There are fried boards, water leaks, software issues, and a host of other problems. Murphy's Law is in full effect today.

During the qualifying heats, most of the vehicles have problems finishing any tasks, and only a handful complete the entire course.

Come the finals on Sunday, the favored MIT

vehicle strays off course by following the reflection of the sun, taking third, while University of Florida's *Subjugator* — a clear cylinder with a rear spoiler and foam strapped on for buoyancy — surprises everyone by taking first, with École de Technologie Supérieure's lunar-lander-looking sub winning the second place slot.

> ❝One reason for the competition is to bring together these bright engineers to create robotic vehicles that can remove human beings from dull, dirty, and dangerous work, such as bomb disposal or de-mining a shipping lane. ❞

Despite the friendly competition and camaraderie between the teams, the potential for heavy-duty military and industrial applications is ever present in the minds of both the students and the competition organizers.

Davidson is clear that one reason for the competition is to bring together these bright engineers to create robotic vehicles that can remove human beings from "dull, dirty, and dangerous work," such as bomb disposal or de-mining a shipping lane.

And he sees the possibilities as endless. "You can design these things to go as deep as the ocean is," he says. "They don't have that human factor of designing a very huge, robust, and very tight system to protect a person. You just pack it all in and you're not constrained by having to protect human life. The same thing goes with the unmanned airplanes. When you take out the cockpit and safety elements, you can pull all sorts of Gs and do things that no manned aircraft could ever do."

Davidson adds, "If you can take people out of the missions that have a risk of life and put a machine in that same mission and something goes wrong, who cares? You blew up a piece of machinery."

Larry Harmon has been covering San Diego music, strangeness, and crime in his fanzine, *Genetic Disorder*, and a number of other publications for more than 10 years.

1. The MIT team's *Orca*, which was favored to win this year's International Autonomous Underwater Vehicle Competition in San Diego, came in third, losing out to the University of Florida's *Subjugator* and École de Technologie Supérieure's *S.O.N.I.A.*

2. Southern Polytechnic State University's Auvrobot landed squarely in the middle of the pack, coming in at 10th place.

3. Because of a last-minute operational mishap, North Carolina State University's delta-shaped *SeaWolf* came in 18th out of 19 entrants.

4. Best in Show: University of Florida's *Subjugator* brought home the gold by successfully completing one of three tasks, and for weighing in at only 30lbs. (far under the 70lb. competition limit).

BACKYARD ZIP LINE

BE THE HIT OF THE NEIGHBORHOOD WITH A HIGH-FLYING TREE-TO-TREE TRANSPORTER.

By Dave Mabe

The author's wife, Joan Nesbit Mabe, tries out the zip line.

YOU COULD BUY A DINKY, READY-MADE kit with a short zip line for kids, but why not make your own industrial-strength zip line that will support the heaviest of neighbors? It's a fun project you can tackle in a weekend. You can order all the parts on the web for less than $300.

First, you'll need to find a suitable location for your zip line. Depending on the lay of the land, you'll be choosing between two basic types of zip line. If you would like to put your zip line on a steep hill, you'll need to use a braked zip line, which has a brake block attached to bungee cords that slow you down as you approach the end of the line. If your site has a more gradual incline, you can use a gravity-stop zip line that simply uses gravity to slow you down. I knew small children would be riding mine, so I chose a gravity stop because it is tamer, and our property layout made this option ideal.

FIRST STEPS

I surveyed my property and identified two large oak trees that were far enough apart and only had a few very small trees between them. I measured the distance between the trees and ordered the following parts from Starlight Outdoor Education (starlightoutdoored.com).

For a permanent zip line, you'll need long eye-bolts, drilled through the entire tree trunk, to attach the cable to. I knew I might move the zip line in the future, so I chose a more temporary technique to attach the cable.

Depending on the run you've selected, you'll need to determine how much cable to purchase. The cable comes in a spool in multiples of 250 feet. You should get more than you need because you'll need a bit of play on both ends.

While I was waiting for the supplies to ship, I started preparing the site. I had to clear some small trees and underbrush to create a path between the two trees. I also started visualizing how far up each tree I was going to need to attach the cable and how steep the slope should be. I knew there would probably be some trial and error, but taking some time to plan ahead of time definitely minimized

this once the supplies arrived.

You'll need a buffer between the cable and the tree it is attached to — otherwise, the tree will actually grow around the cable and, over time, completely engulf it. For this purpose, I bought three 1x6 boards of pressure-treated deck flooring and cut them into several 1-foot lengths. I went ahead and drove the nails partway into the boards so I wouldn't have to do so once I was at the top of the ladder. I nailed the boards into the tree vertically so that they

> ## " Instead of risking life and limb, I took the cable halfway around the tree and then chose the base of another tree to secure the cable. "

encircled the trunk with small gaps in between them. This allows the cable to attach securely to the tree without actually touching it. On a couple of the boards, I drove a nail about halfway in so the cable would be supported by it and wouldn't slide down the tree.

LAYING CABLE

When the cable arrived, I unwound some from the spool, climbed the ladder and circled the tree one full time, and then secured it with two cable clamps. Depending on the diameter of the tree, you may need a second person and ladder to help, as the cable tends to be unwieldy, especially on top of a ladder.

I unwound the rest of the cable toward the other tree and nailed the boards to the second tree. Because the cable on the second tree is so high, I knew it would be difficult and dangerous to tighten the cable and clamp it at that height. Instead of risking life and limb, I took the cable halfway around the tree and then chose the base of another tree to secure and tighten the cable to. This allowed me to use the come-along (a.k.a. the hand winch) to tighten the cable standing on the ground rather than at the top of an extension ladder.

To tighten the cable, I took it around the tree and put two clamps on the cable without tightening them. I then clamped a loop on the end of the cable so I could attach the come-along to it. Because there was so much tightening that needed to occur

MATERIALS:	COST:
7x19 galvanized aircraft-quality cable (250 feet)	$100
Two-wheel zip pulley	$110
Drop-forged cable clamps (8)	$1.75 each
Come-along (chain puller)	$40
Climbing straps (2-3)	$5
Steel carabiners (2-3)	Around $10 each
1X6 decking boards (2)	$10 each
3" nails (several)	Around $5
1 extension ladder	Try to borrow one!

to raise the cable to the proper height, I captured the gain by tightening the cable clamps and then repositioned the come-along and the loop at the end of the cable. This took several repetitions before the cable was tightened (and therefore raised) to a useful riding height.

● TEST RUNS AND SAFETY CONSIDERATIONS

For the first couple of test rides, I started well shy of the top to make sure I wouldn't get a mouthful of bark if I hit the tree at the bottom of the run. I used a wooden handle from an old ax as a hanging bar and drilled an eyebolt through it, using a carabiner to attach the pulley to the hanging bar. I made more test runs by starting closer and closer to the top tree, carefully observing ground clearance and whether height adjustments needed to be made on the bottom tree to create more or less "gravity" at the end of the line.

Depending on the age, strength, and confidence of the rider, there are a variety of options for attaching the rider to the zip line. A climbing harness is the safest and easiest for younger riders, but it is time-consuming to transfer and adjust the harness between different riders. Older and stronger riders can use a couple of climbing straps as a seat by simply attaching the straps to a carabiner and sitting in the loop. The strongest and most confident riders can simply hang from the wooden bar, although because of the height from the ground at the end of the line, this is discouraged due to safety concerns.

1. Protect the tree using the 1X6 boards (otherwise the tree will grow around the cable).

2. Tighten the cable using the come-along.

3. Attach the pulley to the cable.

Dave Mabe is the author of *BlackBerry Hacks* (O'Reilly Media) and lives in Chapel Hill, N.C.

Microsoft

We Put the Power of Windows Embedded in our Robots

Each day, device developers like Robert Bonin are harnessing the power of Microsoft® Windows Embedded to build great devices, like the latest industrial robots from Chaveriat Robotique.

Microsoft® Windows CE provides hard real-time capability to their robots and powers a user-friendly handheld device which enables factory floor personnel to make program changes on the fly. With these and other innovations based on Windows CE, Chaveriat Robotique has reasserted itself as a serious force in the world of robotic arms, with robots that are more functional and less expensive.

What's more, Windows Embedded offers the timesaving tools, operating system technologies and thousands of drivers device developers need—so that they have the power to truly innovate.

❝ *We chose Windows CE because it offers real-time and graphics at the right price.* **❞**
— ROBERT BONIN **/ Research & Development Manager /** Chaveriat Robotique / France

The Power to Build Great Devices—get it with Windows CE, Windows XP Embedded, or Windows Embedded for Point of Service.

www.learnaboutembedded.com/robots10

 **Windows Embedded**

Make: Projects

Get your sugar rush setting off a soda-bottle rocket powered by air and water. Then harness the wind with a PVC windmill generator made from pure junk. And if that isn't enough excitement for you, raise your pulse by spending an afternoon igniting a high-octane jam jar jet engine.

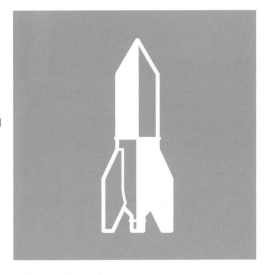

Water Rocket

78

Wind Turbine

90

Jar Jet

102

SODA BOTTLE ROCKET

By Steve Lodefink

You don't have to be Burt Rutan to start your own rocket program. With a few empty soda bottles and some PVC pipe, you can build a high-performance water rocket. >>

Set up: p.82   Make it: p.83   Use it: p.88

LIQUID FUEL ALTERNATIVE

I've been a big fan of model rocketry since I built my first Estes Alpha back in third grade. Nothing is more exciting to a 9-year-old proto-geek than launching a homemade rocket. But flying those one-shot solid-fuel rockets can burn a hole through a young hobbyist's wallet faster than they burn through the atmosphere, and with today's larger, high-powered rockets, locating and traveling to a safe and suitable launch site can require substantial planning and effort.

Instead, you can use 2-liter carbonated drink bottles to build an inexpensive, reusable water rocket. The thrill factor is surprisingly high, and you can fly them all day long for the cost of a little air and water. It's the perfect thing for those times when you just want to head down to the local soccer field and shoot off some rockets!

Steve Lodefink works as an interactive designer and web producer for The Walt Disney Internet Group in Seattle.

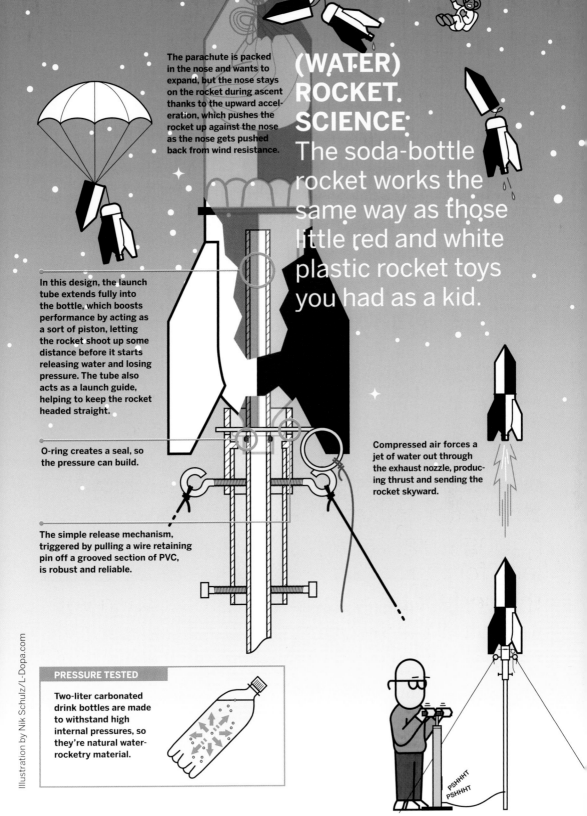

(WATER) ROCKET. SCIENCE.

The soda-bottle rocket works the same way as those little red and white plastic rocket toys you had as a kid.

The parachute is packed in the nose and wants to expand, but the nose stays on the rocket during ascent thanks to the upward acceleration, which pushes the rocket up against the nose as the nose gets pushed back from wind resistance.

In this design, the launch tube extends fully into the bottle, which boosts performance by acting as a sort of piston, letting the rocket shoot up some distance before it starts releasing water and losing pressure. The tube also acts as a launch guide, helping to keep the rocket headed straight.

O-ring creates a seal, so the pressure can build.

The simple release mechanism, triggered by pulling a wire retaining pin off a grooved section of PVC, is robust and reliable.

Compressed air forces a jet of water out through the exhaust nozzle, producing thrust and sending the rocket skyward.

PRESSURE TESTED

Two-liter carbonated drink bottles are made to withstand high internal pressures, so they're natural water-rocketry material.

PSHHHT
PSHHHT

SET UP.

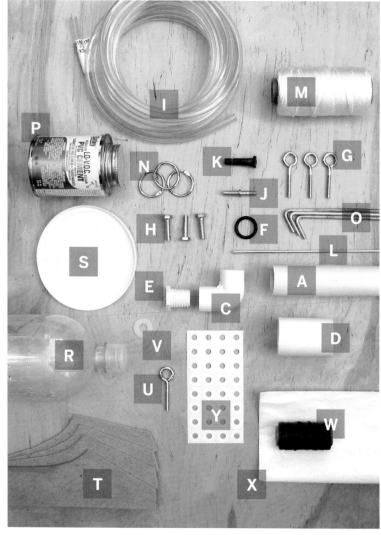

LAUNCHER PARTS

[A] 4" length of 1" PVC pipe **For the release body**

[B] 50" length of ½" Schedule 40 PVC pipe **For the launch tube (not shown)**

[C] ½" PVC elbow pipe **For the end cap**

[D] 1" PVC pipe coupler **For the release collar**

[E] ½" PVC plug cap

[F] Rubber O-ring, 22mm outside diameter (OD)

[G] Eyebolts (3)

[H] Hex bolts (3)

[I] 15' length of ⁵⁄₁₆" OD x ³⁄₁₆" inside diameter (ID) flexible vinyl tubing

[J] ³⁄₁₆" hose barb

[K] Tire air valve

[L] ⅛" music wire **For the release spring**

[M] Nylon cord

[N] Small binder rings (3) **For stay clips**

[O] Small tent stakes (3) **For stays**

[P] PVC cement

[Q] Bicycle pump with pressure gauge **(not shown)**

ROCKET PARTS

[R] 2-liter carbonated drink bottles (3)

[S] 4" deli cup lid

[T] Fin material, such as balsa, thin plywood, or Plastruct sheeting

[U] 2" eyebolt

[V] Medium nylon washer

[W] Kite string

[X] Large garbage bag **For parachute material collar**

[Y] Round hole reinforcement labels

[Z] Quick-set epoxy **(not shown)**

TOOLS

Hacksaw

Utility knife

⅛" file

Drill

Locking pliers

120-grit sandpaper

Thread-cutting taps and dies (optional)

Photograph by Kirk von Rohr

BUILD YOUR SODA-BOTTLE ROCKET

START ⟫⟫   Time: **An Afternoon** Complexity: **Low**

1. BUILD THE LAUNCH TUBE

1a. Cut the launch tube. Use a hacksaw to cut the ½" PVC pipe to length. A 50" tube will make a launcher that's a convenient height for most adults to load from a standing position. The ½" Schedule 40 PVC pipe fits perfectly into the neck of a standard 2-liter soda bottle.

1b. Install the O-ring. Mark the O-ring position by fully inserting the launch tube into the type of bottle that you plan to use for your rockets. Locate the O-ring roughly in the middle of the bottle's neck. Use the edge of a file to cut a channel for the O-ring to occupy. Rotate the launch tube often while you work to maintain an even depth of cut, and be careful not to go too deep. Then slip the O-ring over the launch tube and seat it in the groove.

2. BUILD THE RELEASE MECHANISM

2a. Assemble the release body. Cut a 4" length of 1" PVC pipe and press-fit it into the 1" coupler. Cut squarely and deburr all PVC cuts with 120-grit sandpaper.

Photography by Steve Lodefink

2b. Cut the release spring slots.
Insert your bottle's neck into the
release assembly and determine the
distance of the bottle's neck flange
from the end of the bottle. Mark the
flange location on the 1" pipe coupler
and use the hacksaw to cut a ³⁄₁₆"
long slot on each side. These slots
will hold the retainer/release spring.

2c. Attach bolts. Drill three evenly
spaced holes through the release
collar and release body together, and
thread the three eyebolts into these
holes. Similarly, drill three holes in
the lower release body tube to accept
the three hex bolts.

Optional: Cut threads in these holes
with a tap to accept the hex/eyebolts.
If you won't be tapping them, drill the
holes just undersized, and the bolts
will cut through the PVC just fine. Be
careful not to strip these holes.

**2d. Make the retainer/release
spring.** Bend a piece of ⅛" music wire
1½ turns around a piece of scrap ½"
pipe clamped into a vise. The spring
should be roughly V-shaped.

Make a retainer clip for the spring by
drilling two holes in a scrap of ½" pipe.
The ends of the compressed spring
will fit into these holes. This keeps the
spring closed above the bottle's neck
flange, holding the bottle in place.

Tie a 15' trigger line to the clip. At
launch time, you pull the clip off with
this trigger line, which allows the spring
to open and the rocket to take off.

3. MAKE THE AIR HOSE

3a. Drill a ³⁄₁₆" hole in the center of the threaded ½" end cap, and press in the ³⁄₁₆" barb fitting.

3b. Thread the end cap into the elbow fitting and tighten it with a wrench. Using PVC cement, solvent-weld the elbow to the bottom end of the launch tube. The end cap is tapered, so it should require no Teflon tape or adhesive.

3c. Use a utility knife to strip the rubber from the tire valve to one inch from the end. Insert the valve into one end of the ³⁄₁₆" flexible tubing.

Optional: You can use a die to cut threads into the plain end of the valve stem, and then twist it into the tube.

3d. Push the other end of the air tube onto the barb fitting.

4. SET UP AND TEST THE LAUNCHER

4a. Stake down the stays. The launcher is installed in the field using three stays, each consisting of a 72" length of light nylon cord. Stake one end of each line to the ground, and clip the other end of each stay to the eyebolts on the launcher.

Photograph by Topher Lucas (step 4a)

4b. Pressure-test the launcher. Now is a good time to ensure that all the launcher's connections are airtight. Fill a bottle to the top with water (this way, if the bottle fails this pressure test, it will not explode). Quickly invert the bottle and slip it onto the launcher. A little Vaseline inside the neck will help the bottle make a seal against the O-ring. Squeeze the release spring into the slots in the release collar and clip it in place. Use the bicycle pump to pressurize the system to 70psi. If the pressure holds steady, all is well. Otherwise, fix any leaks and test again.

5. ASSEMBLE THE ROCKET

Water rocket designs range from a simple finned bottle to elaborate six-stage systems with rocket-deployed parachute recovery and on-board video cameras. Ours is a painted single bottle affair with wood fins and parachute recovery. Chute deployment is by the passive "nose cone falls off at apogee" method.

5a. Cut 3 or 4 fins from a light, stiff material such as balsa, thin plywood, or Plastruct sheeting. Roughen the surface of the bottle, where the fins will attach, with some sandpaper and then glue the fins to the bottle with epoxy, or a polyurethane adhesive such as PL Premium. Sand the leading edges smooth.

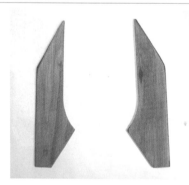

TIP: Gluing on the fins at a slight angle will cause the rocket to spiral as it flies, adding stability to the flight.

5b. Make the nose section by cutting off the neck and base of another bottle. Cut a 6" circle of material from a third bottle. Make a radial slit on the circle, fashion it into a nose cone, and cement it in place atop the nose section.

5c. Outfit the nose cone. When the cement is dry, turn the nose over and epoxy the 2" eyebolt to the inside tip of the nose cone. This bolt serves as a place to anchor the parachute shock cord. It also adds extra mass to the nose section, which will help to pull this section off as the rocket decelerates, exposing the parachute.

5d. Make the nose-stop. Cut the center from a 4" deli container lid, leaving only the outer rim. Cement the rim onto the rocket's lower "motor" section such that it allows the nose to sit loosely and straight on the rocket. This "nose-stop" will prevent the nose from being jammed on too tightly by the force of the launch, which ensures that the nose will separate off and deploy the parachute during descent.

5e. Make a parachute canopy from a 36" or so circle cut from a large trash can liner. For best results, use 12 or more shrouds made from kite string. Apply paper reinforcement labels to both sides of the chute, where the shrouds attach, to keep the chute from tearing. Tie the loose ends of the shrouds to a nylon washer or ring to make the chute easy to manage.

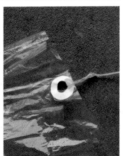

TIP: Ideally, a parachute's shrouds should be a bit longer than the diameter of the chute canopy.

5f. Epoxy a parachute-anchoring ring to the top of the rocket base and tie the parachute to the ring with a short cord. Cut a 4' connecting cord and tie it between the nose cone eyebolt and the parachute-anchoring ring. This cord will keep both halves of the rocket together during descent.

TIP: Make sure the connecting cord is long enough to allow the parachute to completely pull out from the nose cone.

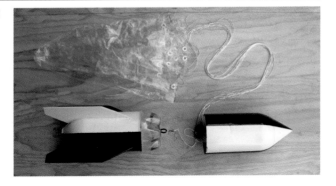

USE IT.

THREE, TWO, ONE, LIFTOFF!

SAFETY

Water rockets produce a considerable amount of thrust, and getting in the way of one could cause severe injury. Take the same common-sense precautions that you would when launching any type of rocket. Make sure that everyone in the area is clear of the rocket and aware that it is about to launch. Do a verbal countdown, or yell something alarming, such as "Fire in the hole!" just before you launch the rocket.

SELECTING A SITE

A well-built, single-stage water rocket is capable of flying several hundred feet into the air and drifting a considerable distance during descent. Less well-built rockets may choose to travel several hundred feet to the side. In any case, you need to choose a launch site that is large and open enough to allow your rocket to wander a bit without getting lost in a tree, or on the roof of some Rottweiler's doghouse.

Big sports fields are the logical site choice for most of us, but if you are in a rural area, any wide-open space will work as a rocketry range. Be sure to take wind direction into consideration when deciding on which side of the field to set up the launcher.

OPERATION

1. Set up the launcher by clipping the three support stays to the launcher's eyebolts. Take up any slack in the lines and stake the other ends to the ground, evenly spaced. Uncoil the air hose and tuck it under one of the tent stakes to keep it from coiling. Attach the bicycle pump to the air hose.

2. Pack the parachute. Grab the center of the parachute canopy between your thumb and fore-finger, and let it hang. Draw the chute through your closed hand to gather it, and then fold it into thirds, zigzag style. Lay down the parachute shrouds on the ground, and accordion-fold them back on themselves. Don't wrap the shrouds around the canopy; just slide the whole thing into the nose section and bring the two halves of the rocket together.

TIP: Line the nose section with parchment paper or Teflon baking sheet liner to help the parachute deploy smoothly. A light dusting of talcum powder will also help keep the chute from sticking.

3. Fill and set up the rocket. While holding the nose in place, turn the rocket over and fill it one-third full of water. Apply petroleum jelly to both the launch tube O-ring and the inside of the bottle's mouth to help it slip onto the O-ring. Hold the mouth of the rocket up to the launch tube. In one smooth motion, pivot the rocket up, slide it down

onto the launch tube, and twist it back and forth, if necessary, to help it engage the O-ring seal.

4. Compress the release spring into the slots of the release collar, locking the bottle flange in place. Install the spring-retaining clip on the ends of the spring, and carefully run the trigger line back to your "ground control" area.

TIP: Tie the release spring to the launcher with some string to keep it from flying across the field and getting lost every time you launch.

5. Launch! Jump over to the bicycle pump and bring the pressure up to about 70psi. When you are ready, clear the area, count down to zero, and pull the trigger line, releasing the spring and freeing the rocket.

If all goes as planned, your rocket will shoot upward, dispensing with its entire fuel load in less than half a second. Then it will begin to decelerate, and the nose will want to separate. As the rocket reaches apogee, the two halves will come apart, deploying the recovery chute and bringing the craft gently back to Earth, much to the excitement of the assembled crowd.

Experiment with different amounts of water and air pressure until you find the sweet spot that sends your rocket the highest. Don't exceed the amount of air pressure that your bottle is designed to withstand; 70psi seems to be about right for a standard 2-liter soda bottle.

The author kindly (and we'd say somewhat naively) allowed the MAKE team to borrow his rocket. After making the rounds through numerous photo shoots, kids' birthday parties, and a high-decibel, code-violating Thanksgiving affair, the rocket sustained the damage you see here. We deeply regret the mishap and promise not to do it to the next rocket we've asked Steve to build for us.

ADVANCED DEVELOPMENT

Once you've tasted the joys of basic water rocketry, you will inevitably want to improve and refine your rocket designs. If you want your rockets to fly higher, the best improvement you can make is to increase the volume of the rocket "motor." This is usually done by splicing or otherwise coupling two or more bottles into a single pressure chamber. There are also various schemes for building multi-stage rockets, as well as more elaborate parachute deployment setups.

There is an abundance of water rocket information available online. Here are a few sources to get you started:

Antigravity Research Corporation – ready-made water rocket components: antigravityresearch.com

Water rocket links:ourworld.compuserve.com/homepages/pagrosse/h2orocketlinksi.htm

The Martinet Launcher, the basis for this project's launcher design: martinet.nl/wp-site/water-rockets

Photography by Topher Lucas

WIND POWERED GENERATOR

By Abe and Josie Connally

With a motor and some piping, it's surprisingly easy to build this inexpensive, efficient wind generator — and enjoy free energy forever. ❯❯

Set up: p.93   **Make it:** p.94   **Use it:** p.101

CURRENT FROM CURRENTS

There are no limits to what you can do with wind power. It's abundant, clean, cheap, and easy to harness. We designed this Chispito Wind Generator (that's Spanish for "little spark") for fast and easy construction. Most of the tools and materials you need to build it can be found in your local hardware shop or junk pile. We recommend that you search your local dump or junkyards for the pieces required. Or, if you live in a city, search freecycle.com for salvaged parts, and see if you can install one on your roof.

We believe that anyone can be in control of where his or her electricity comes from, and there is nothing more rewarding and empowering than making a wind-powered generator from scrap materials. Remember: puro yonke (pure junk) is best!

Abe and Josie Connally are off-grid adventurists based in the remote Big Bend region of Texas, where they experiment and live with sustainable technologies built from puro yonke (*velacreations.com*).

WIND GENERATOR BLOW-BY-BLOW

The Chispito Wind Generator is a simple little machine that's great for getting started with wind power. In a 30 mph wind, ours gives us about 84 watts, 7 amps at 12 volts.

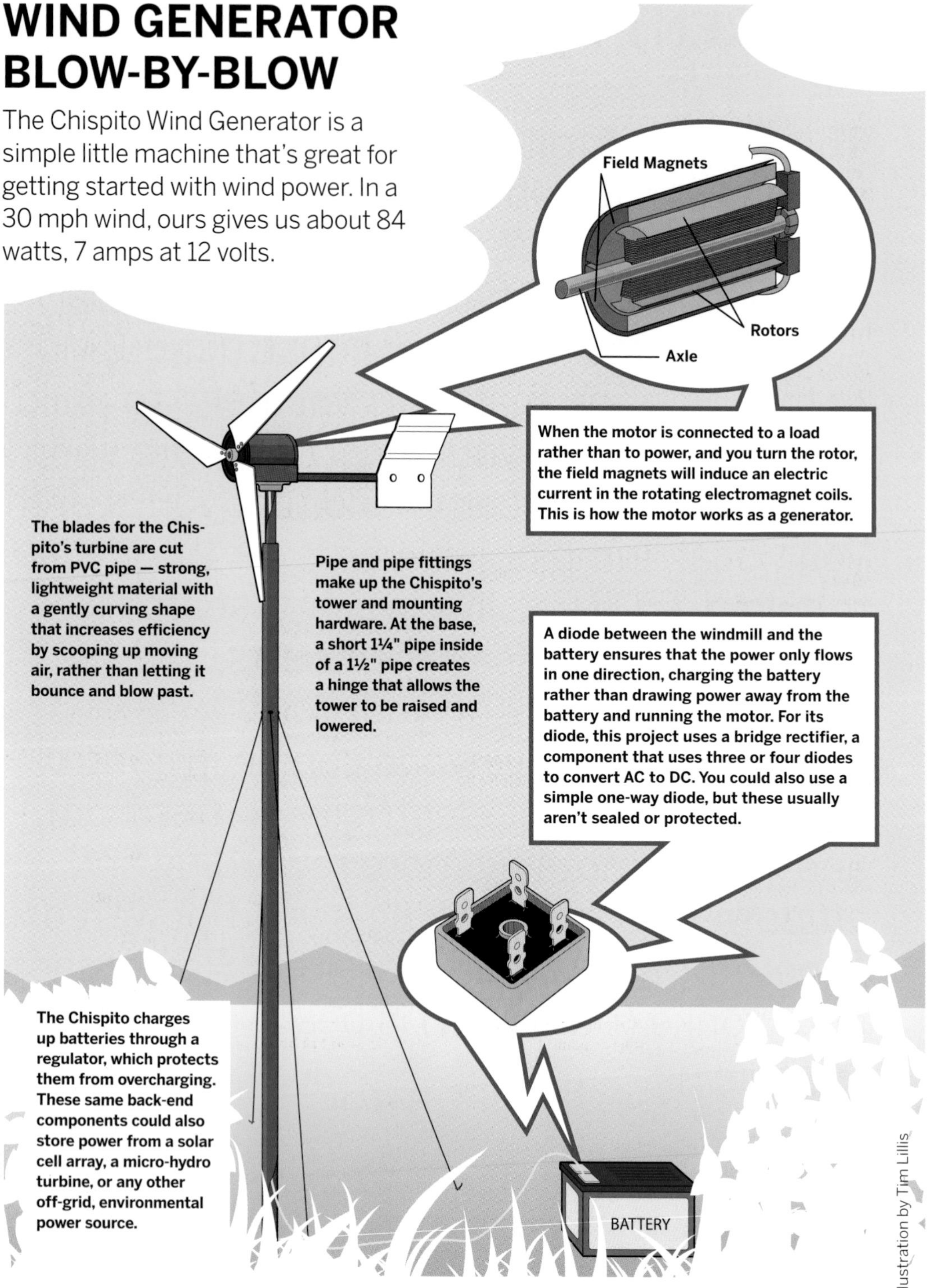

Field Magnets

Rotors

Axle

When the motor is connected to a load rather than to power, and you turn the rotor, the field magnets will induce an electric current in the rotating electromagnet coils. This is how the motor works as a generator.

The blades for the Chispito's turbine are cut from PVC pipe — strong, lightweight material with a gently curving shape that increases efficiency by scooping up moving air, rather than letting it bounce and blow past.

Pipe and pipe fittings make up the Chispito's tower and mounting hardware. At the base, a short 1¼" pipe inside of a 1½" pipe creates a hinge that allows the tower to be raised and lowered.

A diode between the windmill and the battery ensures that the power only flows in one direction, charging the battery rather than drawing power away from the battery and running the motor. For its diode, this project uses a bridge rectifier, a component that uses three or four diodes to convert AC to DC. You could also use a simple one-way diode, but these usually aren't sealed or protected.

The Chispito charges up batteries through a regulator, which protects them from overcharging. These same back-end components could also store power from a solar cell array, a micro-hydro turbine, or any other off-grid, environmental power source.

BATTERY

Illustration by Tim Lillis

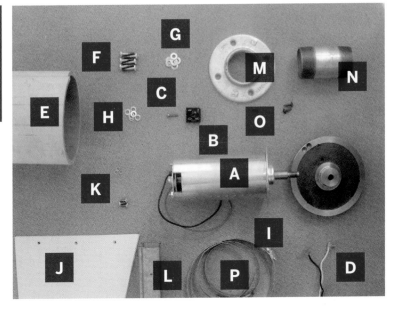

MATERIALS

MOTOR AND ELECTRIC

[A] 260 volts DC, 5 amps, treadmill motor with a 6" threaded flywheel **You may use any other simple, permanent-magnet DC motor that returns at least 1V for every 25 rpm and can handle upwards of 10 amps. Our motor is rated at 5A, no load, and we've found that the coils can withstand 15A going through them without heating up.**

If you use another motor, change this supply list to match. For example, if the motor lacks a flywheel, you will have to find a hub for it. A circular saw blade with a ⅝" shaft adaptor will work.

[B] 30-50A bridge rectifier with center hole mount surpluscenter.com, **item #22-1180**

[C] Mounting screw **(1)**

#8 (or larger) copper wire, red and black, both stranded **Enough length for both a red and black piece to run from the top of the tower, down through length of pole, to batteries. We recommend at least #8 wire, but if your tower will be sited a long distance away from your batteries, you may need a heavier gauge.**

[D] Spade connectors for wires **(4) For bridge rectifier**

Heat-shrink tubing or electrical tape

Battery bank **We recommend deep-cycle lead-acid storage batteries, and a total battery bank capacity of at least 200 amp-hours.**

Ammeter

Regulator or charge controller

Fuse

BLADES

[E] 2' length of 8" Schedule 80 PVC pipe **If PVC is UV resistant, you will not need to paint it.**

[F] ¼" #20 bolts, ¾" long **(6)**

[G] #20 washers **(9)**

[H] Lock washers **(6)**

[I] Hose clamp **(1)**

VANE

[J] 1 sq. ft. (approx.) of sheet metal

[K] Mounting screws and lock washers **(approx. 9)**

MOUNT AT TOP OF TOWER

[L] 36" of 1" square metal tubing or 1" angle iron

[M] 2" floor flange pipe fitting

[N] 2" steel pipe nipple, at least 4" long

[O] Mounting screws **(2)**

MOTOR MOUNT

[P] #72 hose clamps **(2)**

TOWER POLE

10'-30' length of 1½" steel pipe, **threaded at both ends**

TOWER BASE

2'x1¼" steel pipe nipple **(2)**

6"x1¼" steel pipe nipple

1¼" 90-degree steel pipe elbows **(2)**

1½" steel pipe T

10 lb. bags of quick-mix concrete **(2-3)**

¾" #10 sheet metal screws **(4)**

TOWER STABILITY

Guy wire, galvanized steel **With a working load of 200 pounds**

1½" U-bolt

Stakes **(4)**

Turnbuckles **(4)**

TOOLS

Drill and drill bits (⁵⁄₃₂", ⁷⁄₃₂", ¼"), jigsaw, thread-tapping set, pipe wrench, crescent wrench, flathead screwdriver, vise and/or clamp, wire strippers, metal punch or awl, tape measure, level, marker, tape, compass and protractor, shovel, wheelbarrow, several ropes (each at least twice the length of the guy wires), and an extra person or two to help.

MAKE IT.

BUILD YOUR WIND-POWERED GENERATOR

START ⋗⋗   **Time: A Couple Weekends   Complexity: Medium**

1. CUT THE BLADES

Let's begin by cutting.

1a. Place the 24" length of PVC pipe and square tubing (or other straight edge) side by side on a flat surface. Push the pipe tight against the tubing and mark the line along the length of the tube. This is Line A (see Fig. 1).

1b. Starting from Line A, draw parallel lines at 75-degree intervals along the length of the pipe. You should have a total of five lines on your pipe as shown in left Figure 1. Note that one strip will have an arc width of only 60 degrees. That's OK.

1c. Use a jigsaw to cut along the lines, splitting the tube into five strips. Four will be wider than the fifth (60°) strip. Set the 60° strip aside for now.

1d. Place the four 75° strips concave-side-down. For each one, make a mark 20% of the width of the strip from one corner along the diagonally opposite side as shown (see Fig. 2).

1e. Mark a diagonal line between the two marks you just made on each piece, and use the jigsaw to cut along these lines (see Fig. 3). You should wind up with eight identically shaped trapezoidal blades. You can trim a ninth blade out of the 60° strip left over. You now have enough blades for three generators, or plenty of spares for one generator.

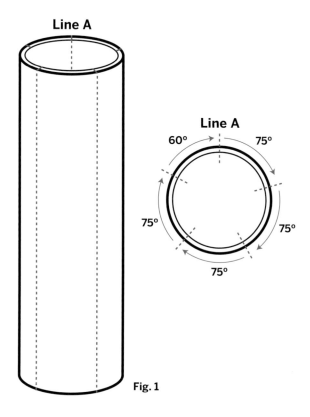

Fig. 1

1f. Now you are going to cut one corner from each blade. First, measure the width of the blade (if you are using an 8" diameter PVC pipe as your stock, it should be about 5.75" wide). Call this value W. Then make a mark along the diagonal edge of the blade, a distance of W/2 from the wide end (3" is good enough if you are using 8" PVC). Make another mark on the wide end of the blade at 15% of W from the long straight edge (1" with 8" PVC) (see Fig. 4).

1g. Connect these two marks and cut along the line. Removing this corner prevents the blades from interfering with each other's wind.

1h. The blades should look like the ones shown in Fig. 5. Pick the three best ones of the batch and let's move to the next step, making the tail.

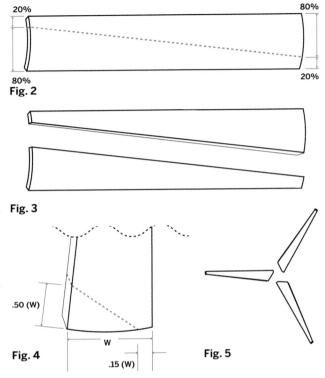

20%   80%

Fig. 2

80%   20%

Fig. 3

Fig. 4

.50 (W)

W

.15 (W)

Fig. 5

---

2. MAKE THE TAIL

2a. Cut the tail. You can make it any shape you want, as long as the end result is stiff rather than floppy. The exact dimensions of the tail are not important, but you'll want to use about one square foot of lightweight material, preferably metal.

2b. Using the 5/32" drill bit, drill two or three holes, spaced evenly, in the front end of the tail. Then place the tail on one end of the square tubing, noting that it will attach to what will become either the right or left side of the tubing, as the generator sits upright. Mark the tubing through the tail holes.

2c. Drill holes in the square tubing at the marks you just made.

2d. Attach the tail to the tube with sheet metal screws. (Or you can do this later, so it doesn't get in the way.)

3. ATTACH THE BLADES

3a. Take three blades. For each blade, mark two holes along the long, right-angle side of the blade (as opposed to the long diagonal side), at the wide end, next to the cut-off corner. The first hole should be ⅜" from the long side and ½" from the end, and the second hole should be ⅜" from the straight edge and 1¼" from the end.

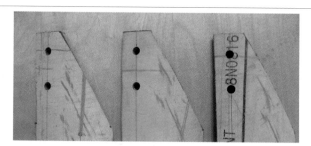

3b. Using the ¼" drill bit, drill these six holes for the three blades.

3c. Detach the hub from the motor shaft. With our motor, we removed the hub by holding the end of the shaft firmly with pliers and turning the hub clockwise. This hub unscrews clockwise, counter to the usual direction, which is why the blades turn counterclockwise.

3d. Using a compass and protractor, make a template of the hub on a piece of paper. Then mark three holes, each of which is 2⅜" from the center of the circle, 120 degrees apart, equidistant from each other.

3e. Place this template over the hub and use a metal punch or awl to punch a starter hole through the paper and onto the hub at each hole.

3f. Drill the holes with the ⁷⁄₃₂" drill bit, then tap them with the ¼" tap.

3g. Attach the blades to the hub using ¼" bolts, running them through the holes closest to the ends of the blades. At this point, the three outer holes on the hub have not been drilled.

3h. Measure the distances between the tips of each blade, and adjust them so that they are all equidistant. Then mark and punch starter holes for the three outer holes on the hub through the empty holes in each blade.

3i. Label the blades and hub so that you can match which blade goes where.

3j. Remove the blades, and drill and tap the three outer holes on the hub.

3k. Position each blade in its place on the hub, so that all the holes line up. Using the ¼" bolts and washers, bolt the blades back onto the hub. For the inner three holes, use two washers per bolt, one on each side of the blade. For the outer holes, just use one washer next to the head of the bolt. Tighten.

4. ASSEMBLE THE GENERATOR

4a. Drill a ⁵/₃₂" hole in the tubing, about 5 inches from the front end of the tube, opposite the tail holes end, on any side. Place the bridge rectifier over the hole, and screw it to the tubing using a #10 sheet metal screw.

4b. Using hose clamps, mount the motor on the end opposite the tail. Do not tighten the clamps, because you will make a balance adjustment later.

4c. Crimp spade connectors onto the black and red wires from the motor, and connect them to the two AC voltages in terminals on the bridge rectifier, L1 and L2. Insulate connections with heat-shrink tubing or electrical tape.

4d. If you haven't already, attach the tail.

4e. Re-attach the blade assembly on the motor.

4f. Now we'll attach the tower mount. Using a pipe wrench, screw the nipple tightly into the floor flange. Clamp the nipple in a vice so the floor flange faces up and is level.

4g. Set the generator on the flange/nipple and balance it by adjusting the position of the motor, then tighten the hose clamps down. Mark spots in the square tubing that match up with the flange holes.

4h. Drill these two holes using a ⁵/₃₂" drill bit. (You will probably have to take off the hub and tail to do this.)

4i. Attach the square tubing to the floor flange with two sheet metal screws.

MAKE THE TOWER ❯❯

5. PLANT THE TOWER BASE

The tower is one of the most important components in your wind generator system. It must be strong, stable, easily raised and lowered, and well anchored.

5a. Dig a round hole about 1 foot in diameter and 2 feet deep.

5b. Feed the 6"x1¼" steel pipe nipple through the horizontal part of the 1½" steel pipe T.

5c. Screw the pipe elbows onto each end of the nipple, one on either side of the T, so that they both point in the same direction.

5d. Screw the two 2'x1¼" pipe nipples into the free ends of the elbows.

5e. Set this hinged base assembly in the hole, so that the T just clears the ground. Dig around, adjust, and position things so that the 2' nipples point straight down and the horizontal part of the T is perfectly level.

5f. With the base properly positioned, mix some concrete and pour it into the hole.

6. ERECT AND STAY THE TOWER

The higher your tower is, the more wind your generator will catch, and the more power it will produce.

6a. Drill a large hole about 1 foot from the bottom of the 10'-30' pipe, for the copper wires to exit.

6b. Screw the pipe into the vertical part of the base's hinged T.

6c. Make four strong, flexible rings out of guy wire, about 5 inches in diameter. For each ring, loop the wire around several turns, and twist it closed.

6d. Place the 1½" U-bolt around the pipe, 3 feet from the top of the pipe. Thread the four wire loops around the U-bolt, and space them evenly around the pipe. Then tighten the nuts of the U-bolt.

6e. Secure a guy wire to each of the loops on the U-bolt. Also loop the ropes (safety ropes) through loops on opposite sides of the pole.

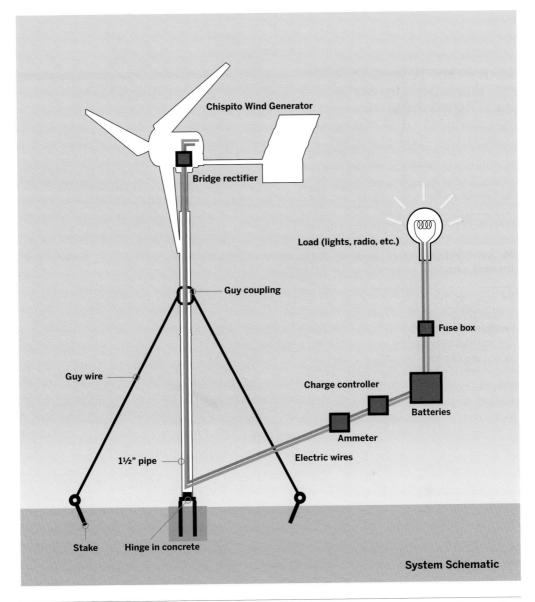

Chispito Wind Generator

Bridge rectifier

Load (lights, radio, etc.)

Guy coupling

Fuse box

Guy wire

Charge controller

Batteries

Ammeter

1½" pipe

Electric wires

Stake

Hinge in concrete

System Schematic

6f. Position the four stakes, spacing them evenly apart at a distance away from the base that's at least 50% of the tower's height. For our 15-foot-tall pole, we positioned the stakes 12 feet away from the base. Then drive the stakes firmly into the ground, slightly angling them away from the base. Or, for greater strength and permanence, dig holes 2 feet into the ground, and set the stakes in concrete.

6g. Wire a turnbuckle to each stake, using several strands of guy wire.

6h. Raise the pole up and tie each of the safety ropes to something solid, like a truck or a building (this is where having another person or two really helps). Attach the guy wires to the turnbuckles.

6i. Hold the pole straight upright, and tighten all turnbuckles to ensure a secure fit.

6j. Mark the front turnbuckle for future reference, so you know how far you need to screw it back in when you're re-raising the pole.

7. WIRE AND MOUNT THE GENERATOR

7a. Release the front guy wire and lower the pole to the ground.

7b. Feed two lengths of #8 wire, red and black, down through the pole and out through the hole in the bottom of the pipe. Then wrap the bottom ends of the two wires together, to create a closed circuit. This is a safety precaution; it puts a load on the wind generator to prevent it from spinning around fast while you're working on it.

7c. Slide the generator assembly over the top of the pole.

7d. Pull the pole wires up through the mount, strip the ends, and crimp them into spade connectors. Plug the red wire into the DC+ terminal of the spade connector (which will probably be perpendicular to the others), and the black wire into the DC− terminal. Insulate connections with heat-shrink tubing or electrical tape.

7e. Raise the pole by pulling the front guy wire into place, and tighten the turnbuckle to the mark made earlier.

7f. Unwrap the ends of the wires and wire up your system as shown in the schematic (previous page). Connect a regulator, an ammeter, a fuse, and a stop switch on the positive line coming from the generator, between the generator and the battery bank. Refer to the manufacturer's instructions. Then hook up the battery bank, and watch it fill up with free power!

Teach a Man to Make

Give a man a fish, or teach a man to fish? Abe and Josie Connally prefer to teach people how to make their wind generators, rather than providing the finished product. They take their wind generators to some of the poorest rural communities in northern Mexico, where rough terrain and long distances make grid power not a viable option. They also hold workshops in their construction for entire villages and have also set up solar water pumps in a few communities.

"Basically, we focus on getting communities to help themselves," Abe says. "So far, we've been met with outstanding enthusiasm."

For more information about their work, and to get complete how-tos of their projects, visit velacreations.com.

"The simplicity and cost are the driving factors. Self-repair and low-maintenance options are essential. Our goal is simple: help those who want to know how to do it."

USE IT.

FREE WIND POWER IS YOURS

SAFETY

Safety should be your highest priority. Human life is more important than electricity, so please follow any and every safety guideline you come across. Wind generators can be dangerous, with fast-moving parts and high voltage, especially in violent weather conditions. Some things to consider:

» Ground and fuse your electrical system, as well as each component within it.

» Always stand upwind when inspecting the wind generator, to avoid debris in case of failure.

» Always attach safety ropes and/or cables when erecting your tower and/or wind generator.

» Always wire connections securely, with proper insulation such as heat-shrink tubing and/or electrical tape.

» Never touch the positive and negative wires at the same time while they are connected to the battery.

» Never leave your wind generator unconnected to anything, unless it is on the ground. Always connect it to a battery or some other load, or else short it out by crossing the positive and negative wires that lead from the generator, to create a closed circuit. Without one of these precautions, the generator can spin freely and attain dangerous speeds.

» Never expose batteries to heat, sparks, or flames, and do not smoke near them. Batteries can ignite and explode easily, resulting in injury.

SYSTEM CARE

To give your wind generator a longer life span, you should paint the blades, and tape the motor with something like aluminum tape.

Batteries are usually the most expensive component in your system, and it pays to take good care of them. The Chispito Wind Generator is rated at 12V, so your system should be configured to match. We recommend deep-cycle, lead-acid storage batteries, and a total bank capacity of at least 200Ah. Check

your batteries' water levels regularly to ensure the longest possible life span. Don't use car or marine-type batteries; they will wear out fast and can be dangerous if used in a battery bank.

For proper system health, you should include a regulator or charge controller, and an ammeter. An ammeter lets you view your wind generator's current-charging rate, and the regulator prevents your batteries from overcharging.

SPECIFIC USES

We use the Chispito Wind Generator, along with a 100-watt solar panel, for all of our electricity needs in our home. This includes a laptop, satellite receiver, TV, VCR, DVD player, printer, stereo, power tools, and lights. Our system consists of a 450Ah battery bank, a 1,000-watt inverter, and the proper fuses and breakers for a home supply.

The wind generator has a wide range of applications depending on your particular needs, from powering your entire house, to running your computer and printer, to recharging the battery packs in your RV. The really nice thing about our model is that you can literally build ten of them for the price of one pre-built wind generator sold in stores.

RESOURCES

Wind generator information and inspiration: otherpower.com

U.S. Wind Energy Resource Atlas wind maps, for assessing the suitability of a region for wind energy: rredc.nrel.gov/wind/pubs/atlas

Edwin Lenz's wind generator made from an old microwave: windstuffnow.com/main/microwave_wind_generator.htm

THE JAM JAR JET

By William Gurstelle

Don't think you can build a jet engine at home? Here's a simple jet engine — a pulsejet — that you can make out of a jam jar in an afternoon. All it takes is bending some wire and punching a few holes. »

Set up: p.105 Make it: p.106 Use it: p.109

Photography by William Gurstelle

JOIN THE JET SET

Turbojets and fanjets contain hundreds of rotating parts. But the ancestors of these designs, called pulsejets, convert fuel and air into propulsive force by using a fixed geometry of chambers and ducts, with no moving parts. The simplest pulsejet is the Reynst combustor, which uses one opening for both air intake and exhaust.

The pioneering Swiss jet engineer Francois Reynst discovered this combustor as a pyromaniac child. He perforated the lid of a glass jar, put a small amount of alcohol inside, and lit the top. Flames shot out of the hole and then were sucked back into the bottle before being ejected again. This almost-magical process repeated until all of the fuel was expended. Reynst had discovered a jar that literally breathed fire, like St. George's dragon. Our jam jar jet is based on Reynst's discovery.

William Gurstelle enjoys making interesting things that go whoosh then splat. He is the author of *Backyard Ballistics* (2001), *Building Bots* (2002), and *The Art of the Catapult* (2004). Visit backyard-ballistics.com for more information.

HOW IT WORKS

When the fuel and air inside the jar first ignite, the jam jar jet generates a burst of hot gas, raising the internal pressure and pushing the gas out. The exiting gas leaves a slight vacuum behind, and fresh air rushes back into the jar to fill the void. More methanol mixes with the fresh air in the still-hot jar, triggering another combustion. Scientists call the resulting cycle relaxation oscillation.

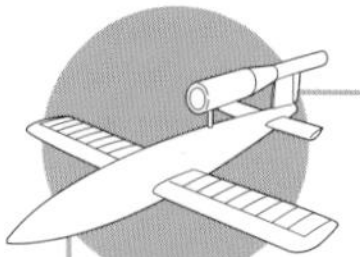

The V-1

The pulsejet engine is simple, cheap, and powerful, but isn't used in commercial aviation because large versions are incredibly noisy, and they vibrate like gigantic, unbalanced chain saws. Invented in the early 20th century, pulsejet engines had no practical use until German scientist Paul Schmidt developed a no-frills but dependable pulsejet-powered cruise missile. This was the notorious V-1 rocket, a.k.a. the "buzz bomb," which terrorized Britain during World War II.

The V-1 was a 25-foot tube with two stubby wings. Its simple engine gave the missile a range of about 150 miles with an explosive payload of nearly a ton. The first V-1 hit London on June 12, 1943. At the height of their use, 190 were launched daily. The V-1 attacks ended only when the Allies marched back through Europe, and seized the missiles' launch sites, which were located across the English Channel.

INTAKE AND EXHAUST

A ½" diameter hole drilled in the center of the Mason jar lid serves as both the air intake and exhaust port. Most functional pulsejet engine designs use two separate ports, but because the combustion cycle's intake and exhaust stages are not simultaneous, pulsejets can also use a single port. The continuous combustion cycles of more advanced jet engines, such as turbojets, require separate intake and exhaust ports.

HEAT DIFFUSION

The copper diffuser also conducts heat and transfers it out to the four wires that it hangs from. The long wires radiate heat to the air outside, which takes some thermal expansion strain off the jar, reducing the risk of cracking the glass.

AIR FLOW

Inside the jar, a conical copper diffuser ring guides the flow of the gases so that they follow a simple whirl pattern. This improves the efficiency of the combustion cycle.

Illustration by Tim Lillis

SET UP.

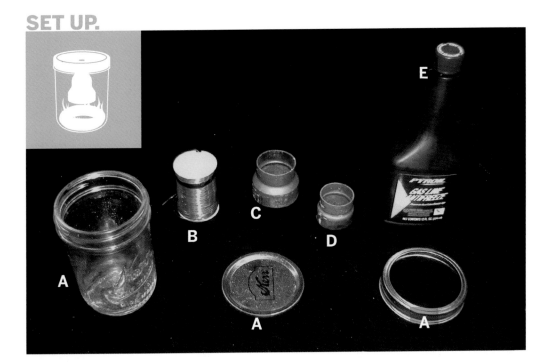

MATERIALS

[A] Pint-sized Mason jar with extra screw caps and lids

[B] 22- to 26-gauge magnet wire (thin enamel-coated copper)

[C] 1½" to 1¼" copper drain/waste/vent (DWV) reducing fitting

[D] 1¼" to 1" copper DWV reducing fitting Available at home centers or hardware stores. These two pipe fittings are for the conical air diffuser.

[E] Small bottle of methanol Available at auto supply stores as gas-line antifreeze; common brands include Heet and Pyroil. Methanol absorbs water readily, which is why it works well as gas-line antifreeze. But this property also causes it to go bad quickly, so you should always use fresh methanol.

[NOT SHOWN]
Package of long fireplace matches, or a long-handled barbeque lighter

Table salt (optional)

Boric acid crystals (optional)

1' long, 1" diameter plastic or metal pipe (optional) **Optional items are for experimental variations.**

TOOLS

Electric drill with ½" and ⅛" drill bits

Wire cutters

File or sandpaper

Teaspoon measure

Cookie sheet

Refrigerator/freezer

Safety glasses

Gloves

SAFETY GUIDELINES

This is a jet engine you're building, a tempest in a teapot. I've never had any problems with this design, but no one — not me, not this magazine — can guarantee your safety. If you do choose to go forward with this project, here are some important safety measures.
1. Do not experiment with different sized jars and openings. A too-large jar with a too-small opening might result in an explosion of glass shards.
2. Use no more and no less fuel than directed. Wipe up any spilled fuel immediately.
3. Use only the parts listed in the directions. These are proven to work safely, and I haven't tried or analyzed all the substitutions that people might think of.
4. Wear gloves and safety glasses or goggles.
5. Do not handle the jar for 5 minutes after a successful run, and then be sure to tap it first to make sure it is cool enough. The Reynst combustor is an extremely efficient heating device, and it gets hot enough to burn skin after just a few seconds of run time.
6. After a long run, the glass jar may crack. If so, carefully sweep the entire assembly into a bag without touching it. Seal the bag and throw it in the trash. Jars are cheap enough, so just get another one.
7. Keep spectators at a safe distance.
8. Always ignite the engine with a long-handled match or barbeque lighter, to avoid getting burned by the pulse of hot gas that immediately follows ignition.
9. Examine all parts for wear before and after use. Discard any worn parts.
10. Always use common sense before, during, and after running the jam jar jet.

MAKE IT.

CONSTRUCT YOUR JAM JAR JET

START »

Time: An Afternoon Complexity: Low

1. DRILL THE PORT

Drill a ½" diameter hole in the lid of the jar. Use a file or sandpaper to completely remove the burr. If the hole is so jagged that it cannot be made smooth and round, discard the lid and re-drill another one.

2. DRILL THE DIFFUSER HOLES

Drill four ⅛" diameter holes in the small copper adapter. The holes should be located about ¼" down from the smaller, 1" diameter end. Space the holes evenly around the perimeter at 90, 180, and 270 degrees from the first hole.

3. ASSEMBLE THE DIFFUSER

Insert the large end of the small copper adapter into the small end of the large copper adapter. Press-fit them together firmly. This forms the conically shaped jet diffuser and heat sink.

4. CUT THE DIFFUSER WIRES

Cut four 4" long wires from the spool of magnet wire.

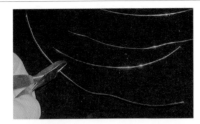

5. ATTACH THE WIRES

Loop one wire through each one of the holes you just drilled, and tie a knot. Extend the other end of the wires outward, radially, from the diffuser cone.

6. SUSPEND THE DIFFUSER

Center the copper diffuser in the middle of the jar. Crimp the wires over the edge of the jar so that the cone hangs suspended close to the top of the jar, with a gap of about ¼" between the diffuser and the top.

7. ADD THE FUEL

Carefully pour or use an eyedropper to measure and add 5 to 10ml (roughly 1-2 teaspoons) of methanol into the bottom of the jar. You can vary the amount of methanol by a small amount to improve performance. At most, the methanol should just cover the bottom of the jar.

8. CLOSE THE JAR

Screw the Mason jar lid down onto the jar and over the copper wires. The lid will hold the diffuser cone securely in place at the top of the jar.

9. VAPORIZE SOME OF THE FUEL

Prepare the jar by letting it sit in the freezer for two minutes. Hold your thumb over the opening in the lid. Vigorously swirl and shake the methanol inside the jar. Place the jam jar jet on a cookie sheet and place the cookie sheet on a secure surface, away from any flammable objects.

NOTE: When you remove your finger from the hole, you should notice a slight pressure release, and the jar should make a very faint "pffft" sound. If you feel no slight pressure and hear no sound, shake the jar again. If there is still no pressure, there is a leak in the seal of the jar that you'll need to fix.

10. FIRE IT UP

Wearing safety glasses and gloves, hold a flame over the opening in the jar's lid.

The fuel will ignite, and for the next 5 to 15 seconds, the jam jar jet will cycle, pulse, and buzz, running at a low but audible frequency of about 20Hz, depending on conditions in the jar and in the surrounding air. With the lights down low, you'll enjoy a noisy, deep blue pulse of flame that grows and shrinks under the lid as the jar breathes fire. It's an amazing effect.

Pint-Sized Fireworks

During the air-intake part of the cycle, the bottom of the jam jar jet glows brightly. The photo on page 102 shows the blue flame you'll get from burning straight methanol, and this photo (at right) shows the yellow variant that comes from adding a little salt to the fuel. By adding salt or boric acid crystals, you can color your flames in a variety of attractive, retina-burning hues, as described on the next page.

TIPS AND TRICKS FOR YOUR JAM JAR JET

FLAMBÉ RECIPES

Here are some interesting variations on the jam jar jet that you can experiment with:

1. For a bright yellow flame instead of the blue, add a pinch of table salt to the methanol.

2. For green-colored flame, add a pinch of boric acid crystals to the methanol.

3. To amplify the sound of the jet, hold a tube a half inch or so above the hole. You can use a metal or plastic pipe, and even the cardboard from a roll of paper towels will last a little while.

Use pliers or a gloved hand to hold the tube in position. Experiment with the length and diameter of the tube. When the size is right, you'll be rewarded with an unmistakably loud, deep, resonant buzz.

4. Some enthusiasts make Reynst combustors with metal jars instead of glass, and outfit them with resonator tubes permanently attached above the hole. These are sometimes termed "snorkelers."

The most advanced snorkelers also have fuel-feed systems that drip methanol into the combustion chamber, which allows them to sustain combustion for long periods of time.

TROUBLESHOOTING

If the methanol burns with a single big whoosh instead of pulsing:

》 Check the size of the hole and make sure it is accurately drilled to a ½" diameter.

》 Be sure to place the jar in the freezer for two minutes before lighting. In my experience, slightly cooling the fuel and the jar improves performance.

》 Make certain the jar is charged with the recommended amount and type of fuel.

If you hold the long match over the opening and it doesn't ignite, or it does ignite but the pulse is weak:

》 Make sure the methanol is fresh.

》 Cool down the jar in the freezer for two minutes.

》 Start with just one teaspoonful of fuel in the bottom, and vary the amount slightly until you get better performance results.

》 Check the seal by listening for the "pffft" when you remove your finger from the hole. If necessary, rejigger the lid to get a good seal.

》 Reposition the diffuser by adjusting the support wires, or try shortening the diffuser by removing the bottom section.

If the jar cracks:

》 Carefully dispose of the broken jar and replace it with another one of the same size. The Reynst combustor/pulsejet is a very efficient burner and therefore extracts a lot of heat from the fuel very quickly. If the jar you're using cannot handle the rapid expansion, it will crack.

RESOURCES

Pulsating Combustion: The Collected Works of F.H. Reynst, Pergamon Press, 1961

Homemade pulsejets webpage and discussion forum: pulse-jets.com

Larry Cottrill's *jetZILLA*, an online magazine of amateur jet propulsion: jetzilla.com

Holes, Rivets, and Bent Metal

Learn three essential shop fabrication techniques and reward yourself by making a wi-fi signal deflector.

How to Drill Equidistant Holes in a Straight Line

Pegboard is underrated. Handy as can be. Think you are too good for pegboard? Julia Child's kitchen at the Smithsonian is practically a pegboard temple.

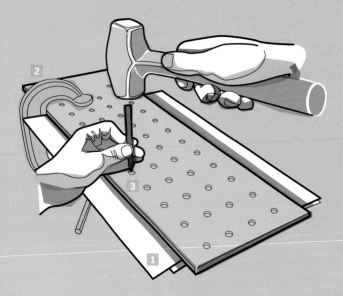

1 Clamp two sheets of sheet metal or plywood to extend over the edge of the bench.

2 Align, square, and clamp a strip of pegboard over the sheets.

3 Use a pointy center punch to mark each hole. Every pegboard hole too many? Skip every other hole. Want staggered holes? Then stagger them!

4 Remove pegboard and drill holes.

> **WARNING:** Sheet metal wants to turn with the drill bit. Keep your fingers by always clamping the metal steadfast!

Drillin' and Snippin'

Drill Big Holes
The Irwin High Speed Metal Unibit step drill bit is the best way to drill holes up to about 1½".

Drill Bigger Holes
What about big holes? Doorknob-size holes? Hole saws can drill holes up to 6". Buy a mandrel and the appropriate hole saws to Swiss-cheese your world.

Best Sheet Metal Shears for the Money
According to legendary custom car builder Gene Winfield, there is no better tin snip for the money than the Irwin Aviation Snip from Harbor Freight. Either the right or left snips will cut a straight line, but if you want to cut a left-leaning radius, you need the left snips. The same goes for a right-leaning radius.

What about all that leftover pegboard?

Hang it on the wall and get organized! A pegboard hanging hardware kit is under $2 and is adequate if you are planning to hang spatulas and wooden spoons. If you want hammers with outlines showing which goes where, screw a pegboard-size 1"x2" pine frame to the wall studs and attach the pegboard.

Illustrations by Damien Scogin

How to Pop-Rivet

There is nothing that looks faster than pop-riveted aluminum. From race cars to airplanes, the pop rivet is the fastener of choice for joining sheet metal. It is a "blind fastener," meaning you don't need access to the backside, as you would if you were using a nut and bolt.

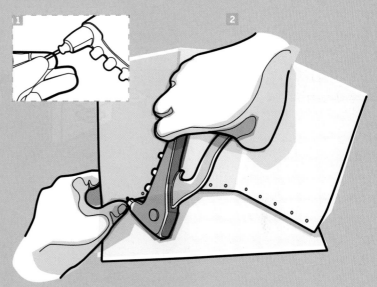

1 Load the right pop rivet for the job (i.e. hole size and sheet metal thickness). Squeeze the riveter handle just until you feel resistance.

2 Put the fat end of the rivet in the hole and squeeze the handles several times with great pressure. The rivet will pop when the pin has collapsed the fat end and finally snaps off due to your awesome brute strength.

HOW A POP RIVET WORKS

How to Bend Sheet Metal

The best way to bend sheet metal is on a sheet metal brake, but you can get decent results with a couple pieces of angle iron clamped to the edge of a sturdy table.

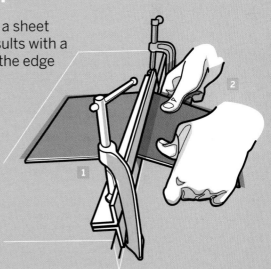

1 Sandwich the piece of material between the angle iron, and clamp tight.

2 Bend to the desired angle!

Tip: Aluminum is easier to bend than steel, but even aluminum can be tough. If the grade is prefaced with a T (e.g., T6, T8, etc.), it is tempered for strength and can crack when bent. The .025" thick 3031 aluminum is swell for most home projects.

How to Build a Power Tap

Build a power tap and put the power where you need it. Electrical junction box systems are the Lego of electrical. All sorts of adapters, connectors, and doodads are available to fit any situation.

MATERIALS
Electrical outlet and face plate
Single-gang junction box
Combination connector
Cheap extension cord

1 Connect a combination connector to the side of the junction box.

2 Feed the cord through the combination connector and tighten so the cord won't inadvertently pull out.

3 Separate and strip the wires. Twist the copper strands and tighten one wire to each side of the electrical outlet — not to the green ground screw!

4 The outlet comes with the correct screws to install it to the junction box.

WARNING: This is safe as an extention cord, but don't wire your whole house like this.

*Please do not attempt to complete this project using the original illustration and instructions on this page (112). They are incorrect and could create a short circuit. Instead, refer to the updated instructions and illustrations online at http://makezine.com/05/quickanddirty/

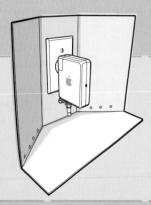

Now Make Yourself Useful

You can use all these tips to make this wi-fi hotspot signal deflector. Put it all together and you end up with a pretty decent wi-fi extender that actually works! In addition to the reflector gain, it really helps to get the Airport higher than a typical wall outlet or a floorbound surge protector. The all-important office-to-garage test netted another bar or two of wi-fi signal strength. Not too shabby considering it can be built for bubblegum money!

Mister Jalopy breaks the unbroken, repairs the irreparable, and explores the mechanical world at Hooptyrides.com.

Sneaky Uses for Everyday Things
By Cy Tymony

Wireless remote alarm sensor: Make different types of alarms and sensors with a cheap remote control car.

You will need: RC car, watch batteries, buzzer, wire, LEDs, seltzer tablet, universal AC adapter.

Inexpensive radio-controlled cars have many sneaky adaptation possibilities. We'll use the cheap single-function type of radio-controlled toy car — the type that continuously travels forward once it's turned on until you actuate its remote control, causing it to back up and turn. You can modify the transmitter to a more compact size, and for use as an alarm trigger. You'll also modify the receiver to activate other devices, such as lights and buzzers.

CHEAP REMOTE
CONTROL CAR

ZOOM

1. The transmitter is always in the Off mode until its activator button is pressed or until wires connected across the push-button leads are connected together.

The first step is to remove the circuit board from the case. Then, substitute lightweight watch batteries for its normal battery type. For each AA or AAA battery, use one small 1½-volt watch battery to supply the same voltage output.

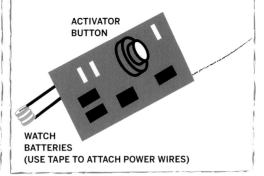

ACTIVATOR
BUTTON

WATCH
BATTERIES
(USE TAPE TO ATTACH POWER WIRES)

PCB IN CAR

2. Remove the radio receiver from the car's chassis. Unlike the transmitter, the receiver must stay On in order to operate. You can use an AC adapter (with the correct output voltage) instead.

The car motor is attached to the receiver with two wires. If you disconnect the motor wires, the receiver can be used for more practical purposes to indicate that a remote sensor is activated. LEDs and buzzers are perfect.

3. If you connect two wires across the transmitter's activator button, you can have another sensor or switch activate the transmitter to alert you of an entry breach or that your valuables are being removed, or that your basement is flooding.

Here, two paper clips are attached to the wires. Then, a seltzer tablet is placed between the paper clips with a rubber band. If water gets near the wires, the tablet will dissolve and the rubber band will cause the paper clips to make contact and activate the transmitter.

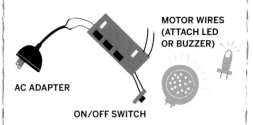

MOTOR WIRES
(ATTACH LED
OR BUZZER)

AC ADAPTER

ON/OFF SWITCH

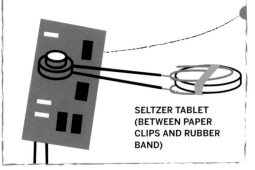

SELTZER TABLET
(BETWEEN PAPER
CLIPS AND RUBBER
BAND)

Cy Tymony (sneakyuses.com) is a Los Angeles-based writer and is the author of *Sneakier Uses for Everyday Things*.

Illustrations by Mark Frauenfelder

The best deal on MAKE guaranteed.

Give your keeper maga-
zines the durability of a
hardcover book.

OLDE-SCHOOL BOOKBINDING

Pages last longer, lie flatter, and look better inside a handsome, durable hardcover.
By Brian Sawyer

Magazines aren't really built to last, but here's how you can turn your copy of MAKE (or any other magazine or printout) into a durable hardcover that will withstand the test of time. Your hardcover MAKE will also lie flat on your workbench, making it easier to follow instructions for other projects.

Create the Signatures

Peel away the existing cover, and use a utility knife and a heavy ruler to cut all the pages out ⅛" from the spine, freeing them from the glue.

Divide your loose pages into consecutive 32-page sections (signatures). Binding loose pages as joined signatures will strengthen the spine and keep pages from falling out. For MAKE Vol. 01, without the ads, I got 192 pages, or six groups. If your total page count isn't a multiple of 32, you can fudge the signature sizes a bit, but each signature must have a page count divisible by 4.

Open the first group in half, such that pages 16 and 17 are facing. Pair these pages and set them aside, doing the same for the next facing pages (14 and 19) and the rest in the group. Just keep

MATERIALS:

Magazine Or pages to be bound	Loose-weave, white linen fabric
Acid-free nontoxic adhesive I'm a Yes man, myself	**TOOLS:**
	Heavy-duty ruler or carpenter's square
Bookbinding tape, ¼" linen	Utility knife
Binder's thread Durable, acid-free linen	Pen
	Binder's needle Or tapestry or other heavy needle
Decorative paper For cover and end sheets	Brush
	Scrap paper
Binder's board, ⅛" thick Or use chipboard or illustration board	Wax paper
	Medium-grit sandpaper
	Medium duty awl

subtracting 2 from the left side and adding 2 to the right to determine which sheets to pair up. If you did it correctly, the last two pages you pair up will be 2 and 31.

Draw a vertical line ¼" from the right edge of page 16 (and all up-facing even pages). Cover the area to the left of the line with scrap paper. Brush a thin coat of acid-free, nontoxic adhesive (such as Yes) into the exposed ¼" gulley. Press the corresponding up-facing odd pages into the gully to glue together the pairs.

Return the pairs to the order they were in before being split in half. Align the edges of the pages and fold them along the spine. Repeat for each group and collate the finished signatures in their original order.

Stitch the Signatures

Sewing your signatures together around bookbinding tape creates the added durability of a hand-bound book. Measure and make two marks along the fold of the first signature, ½" in from each edge of the signature. These two marks represent the kettle stitches, the stitches that connect one signature to the next.

Now measure and mark six more points on the fold: three pairs of points, ¼" apart, spaced evenly between the two kettle stitches. These represent the in- and out-point for sewing the signatures around three tapes, which will run behind the signatures.

Stack the remaining signatures and make the same marks, at the same measurements. Pierce the marks with an awl, making holes just wide enough to allow a needle to pass through snugly.

Using a heavy needle, enter the spine and pull about 30" of thread through the foot (bottom edge) kettle stitch of the last signature. Exit the spine at the next hole and re-enter around the first tape.

Keep stitching around the tapes, and exit the spine at the head (top edge) kettle stitch. Using the same thread, enter the next-to-last signature at its head kettle stitch. Stitch around the tapes, and knot the thread around the foot kettle stitch of the first signature.

Continue in this fashion to stitch the remaining signatures together. If you run out of thread, knot a new 30" length to the existing thread. The best place to do this is just before re-entering the spine around a tape. When you come to the last kettle stitch, knot the thread.

Use a heavy needle to penetrate the paper. The three tapes (one is shown here) run behind the signatures.

At this point, it's a good idea to apply a bit of glue (about ¼") to the inside of the first and last signatures (use a piece of scrap paper to protect the portion of the page you don't intend to glue), and put the work under heavy weights overnight.

Glue the Spine

Next, attach the mull (a strong strip of cloth with a loose weave that allows paste through it) to the spine and tapes. (Connecting the cover boards to the mull, rather than directly to the signatures, allows for a flexible backbone. This is the key to lay-flat binding.) Keeping the pages aligned on all sides, sandwich your work in a press or vise.

Cut a piece of mull that's tall enough to cover your kettle stitches, and 3 inches wider than the width of the spine. Brush a generous amount of

1.

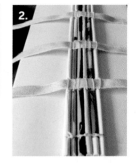

2.

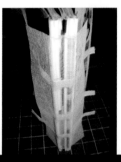

From magazine to book: Attaching the signatures with three strips of bookbinding tape (1), gluing the mull (linen fabric) to the spine (2), gluing on the cover boards (3), and smoothing out cover wrinkles (4).

To see a complete set of annotated photos of the author's bookbinding process, you can visit flickr.com/photos/olivepress/sets/14697/.

3.

4.

Ready to sew. Though using a stitching post is not absolutely necessary (as long as you keep a steady hand, make sure the tapes remain taut, and ensure that the signatures stay even while you sew), it does keep the work organized and easier to manipulate with the only two hands you have. If you have one available, set it up with the tapes stretched tight and spaced to match your holes.

glue on the spine, from kettle stitch to kettle stitch, across the full width. Lay the mull flat and mark the spine area on it, 1" in from either side. Brush this area generously with glue.

Position the mull symmetrically and rub it into the spine. Leave the book overnight under heavy weights or in a press.

Attach Cover Boards

Cut three boards (⅛" binder's board) for the front cover, back cover, and spine. Allow ⅛" additional clearance on the head, foot, and fore edges. Reduce the edge on the spine side of the front and back covers by ¼" (the thickness of two boards), to accommodate the hinge. Altogether, the covers should be cut to ¼" taller than the height of the book (⅛" added to the head and foot) and ⅛" narrower than the width of the book (adding ⅛" and subtracting ¼").

Cut the spine to the same height as the covers and the same width as the signatures. Sand down the rough edges.

Place one piece of wax paper between the mull and the free ends of the tapes and another beneath the tapes. Brush the free mull edge with glue. Remove the top piece of wax paper. Press the front board against the mull, extending ⅛" of the board over the head, foot, and fore edges.

Open the cover and rest it against a board for support. Rub the mull with cloth or paper, working the glue into the board. Brush the tapes with glue and press them to the cover board. Discard the second piece of wax paper, and place another clean piece between the cover board and the first signature. Repeat for the back cover.

Cover the Cover

You'll now cover this skeleton with a single piece of decorative paper, which wraps around the front cover, back cover, and spine. To create room to slide the paper over the edges of the cover and spine, use your utility knife to slit the mull by ½" where the covers meet the spine, at both the head and foot edges.

Lay your cover paper face down and mark the placement of your boards. Allow a ½" (thickness of four boards) turnover width for all edges, and ¼" (thickness of two boards) for each hinge.

Brush a generous amount of glue on the spine, from the head kettle stitches to the foot kettle stitches and the full width of the spine, including the portions of the tapes that rest over the spine (don't paste the free ends of the tapes). Then, attach the mull to the spine, covering the head and foot kettle stitches, and work the cloth into the spine with your fingers and a clean, dry rag until the glue is set.

Brush the spine area of the paper with glue, position the board, and press firmly. Turn the paper over and rub to secure the spine and mold the paper over the edges of the board.

Brush the area you've marked for the front cover with glue, brush slightly into the turnovers and hinge, and press the board between your marks. Turn the book over and rub the cover to remove air bubbles or wrinkles.

Brush glue onto the area for the back cover. Lay the back cover board's fore edge down on the paper, meeting your mark for that edge. Pinch the paper to the board on that edge, and press the remaining paper to the back of the book.

Rub the back cover, working the paper on into the hinge to seal the paper to your book block around the spine. Repeat for the hinge of the front cover.

Lay the book open, and brush glue across the length of the head turnover. Stand the book up on its foot edge and roll the edges of the head turnover over the top of the board. Repeat for the foot edge. Brush fore edges with glue, fold the turnovers, and smooth out wrinkles.

Lay sheets of wax paper between the cover boards and your block of signatures, and press under heavy weights overnight.

Finish Up

Pasting end sheets to the inside front and back covers reinforces the spine and finishes your book's appearance. I used the original MAKE wraparound cover as my front end sheet. Trim it to leave an equal distance around each edge, and paste it to the board. Allow glue to run into the spine, covering the point where the cover meets the spine and extending into the first signature by a ¼" gutter. Repeat for the back cover board.

Put fresh pieces of wax paper between the covers and the book block, and then set under heavy weights to dry overnight. You'll wake up to a long-lasting volume that will look unique, lie flat, and serve you well.

Brian Sawyer (bsawyer@oreilly.com) is lead editor for O'Reilly Media's Hacks Series. When he's not under the hood of a paid project or working pro bono for his wife, he can often be found fiddling with books in various capacities.

OUT DAMNED SPOT!
The chemistry of stain removal.
By Arwen O'Reilly

Both absent-minded and a klutz, I have had my fair share of debilitating stains over the years, and cultivated every stain tip I could get my hands on, even trying tricks from old, dusty, out-of-print books. Removing stains doesn't have to be hard; usually it's just a question of knowing the right chemistry (and practice makes perfect).

Different types of stains are more soluble in different temperatures: hot water, for example, dissolves sugars more easily, while it will set many proteins, like blood. Enzyme detergents will break up long-chain molecules, and are more successful at getting rid of protein stains like grass. In a pinch, dishwashing soaps can help, too; they are designed to break down food proteins and grease, after all. At the very worst, you may have to try a solvent. The following are a few of my favorite tricks, old and new. (Always test on a spot that isn't easily noticed first, and wash stains from the back, forcing them off fabric.)

Berries/Jam/Honey
Boiling water, especially when poured from a distance. This sounds like madness, but it is absolutely miraculous. Makes berry picking fun again.

Grease/Oil/Butter
Baby powder applied thickly, left overnight. Brush off powder in the morning (I use a toothbrush to clear the caked powder off the stain). If there's still a mark, reapply powder, and if not, launder.

Red Wine
Pour white wine onto the stain! Another miraculous save. If you don't have white wine, soda water will do. After, pour salt onto the stain to absorb the liquid. Try to get to it as quickly as possible.

Coffee/Tea
Borax and water mixture 3-to-1 (for carpets, use a non-gel shaving cream and scrub with a toothbrush). Glycerin also works magic.

Gum/Wax/Tar
Rub with ice and scrape off once hardened. With wax, if any excess remains, iron a paper bag over it, which will absorb the wax.

Blood
Cold saltwater and soap — not hot!

Avocado/Tomato/Ketchup
Enzyme detergent and a mild bleach (lemon, white vinegar, hydrogen peroxide).

Grass
Rub the back of the stain with rubbing alcohol. Scrub with non-gel baking soda toothpaste.

Sweat
Ammonia or white vinegar, then soak with laundry detergent.

White-out/Crayon/Lipstick
Use WD-40 and rinse, or try acetone (nail polish remover). Dishwashing liquid and hot water help, too. For lipstick, try bar soap first.

Ink
Use rubbing alcohol, turpentine, or acetone. Soak in milk or rub with toothpaste.

Stain-Removal Links

Cornell's *Removing Stains at Home* PDF
makezine.com/go/cornellpdf

FabricLink's stain removal site
makezine.com/go/stain2

Butler's Guild stain removal site
makezine.com/go/stain3

Visit makezine.com/05/diy_stains **for more tips.**

Arwen O'Reilly is assistant editor of MAKE.

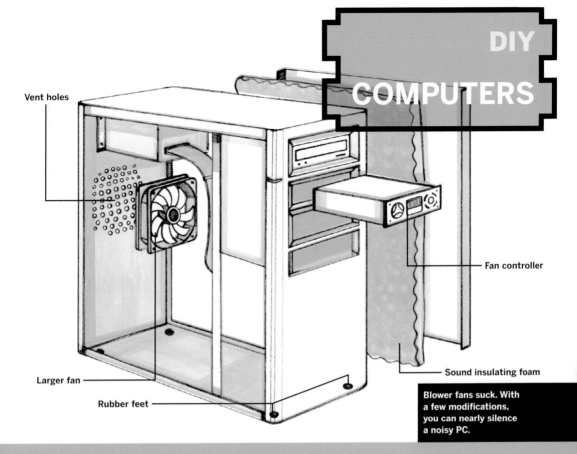

Vent holes

Fan controller

Larger fan

Rubber feet

Sound insulating foam

Blower fans suck. With a few modifications, you can nearly silence a noisy PC.

PUMP DOWN THE VOLUME
Five simple steps to a quieter PC.
By Jason Kohrs

Illustration by Damien Scogin

No one wants their computer to be loud, but in order to keep components running at safe temperatures, cooling fans are often needed. Unfortunately, they can make the system sound like a blow dryer. In a busy office environment, some noise may go unnoticed, but at home, the whir of a fan is annoying.

Silencing a computer can be a costly endeavor, but taking a few relatively inexpensive steps can have a drastic impact on the noise produced by the common computer system. Before starting on any sound reduction upgrades, analyzing a system to pinpoint the areas in need of the most attention will help determine the best course of action and the best way to spend any money.

Here are a few ways to quiet existing systems on a minimal budget.

Cooling Fans
The bulk of all noise in a computer system is going to come from the cooling fans mounted on the case and from any heat-generating components such as the processor. Cases generally employ 80mm fans with ball bearings to keep cool air flowing. Two steps to reduce noise include increasing the fan size and choosing a fan with fluid or sleeve bearings. If a 120mm fan can be installed where the 80mm fan presently resides, a noise reduction can be achieved because the larger fan can move the same amount of air at a lower rotational speed. In general, the slower a fan moves, the less noise it will make.

The ball bearings on many fans are a source of noisy vibration. Selecting a fan with fluid or sleeve bearings will greatly reduce the noise created, which is generally a good thing, except for one instance. Ball bearing fans can be counted on to get even noisier just before failure, letting you know when replacement is necessary. Fluid or sleeve bearings will just fail without such a warning, which could jeopardize other system components.

A quality "processor cooler" (a special add-on fan designed to blow air on a PC's microprocessor) is essential to keep a high-powered system running cool, but it isn't always necessary to run the fan installed at full speed. Some coolers, such as the Cooler Master Aero 4, include a simple speed dial that can be mounted either in the back or the front of the case for convenient adjustment. For those bold enough to run plumbing inside a computer, water cooling kits such as the Cooler Master Aquagate can take cooling performance and quiet operation to a whole new level.

Cases

A case with ample ventilation is required to keep the components cool, and a few things can be done to achieve this without adding to the noise level. Of most interest is the availability of multiple fan-mounting locations in a case, as well as the open area provided for the fans to move air.

Often, the perforations provided for air to pass through are somewhat restrictive, which could add to the noise level as the wind whistles through the small openings. This is nothing that someone handy with a Dremel couldn't remedy.

Experimenting with the size, speed, and placement of case fans can lead to a setup with adequate cooling and low noise levels. It is possible for some cases to be cooled well with a single 120mm exhaust fan, while leaving the other various fan locations empty. There is no need to use all of the fans just because they are there.

Fan Controllers

Fan controllers are available in numerous configurations, but they all serve the same function — to allow a fan to run at something other than full speed. Just reducing a fan's speed by 5-10% can have a noticeable impact on noise, but zero impact on cooling performance.

Some fan controllers operate automatically, using a thermal sensor to vary the speed of the fan in direct proportion to the temperature sensed. This type is convenient as it requires

no user interaction but eliminates any possibility of custom control.

Manual speed controllers put all of the power in the user's hands, generally with a dial that adjusts the fan's speed by varying the resistance on the line powering it. The Cooler Master CoolDrive 4 is primarily a hard drive cooler, but it also serves the function of a four-channel, manual fan-speed controller. From one digital control panel, up to four temperatures can be monitored, and the corresponding fans can be monitored and controlled to maintain a healthy balance between noise and temperature.

Power Supplies

The typical computer power supply features two 80mm fans to keep it cool, which will obviously also generate some noise. Fanless power supplies are now available that generate zero noise, and there are other ways to quiet a power supply without removing the fans altogether. The MGE Vigor 450W Power Supply reduces noise by using a larger 120mm fan to move more air with less speed and a fan speed control knob to allow the user to reduce the speed even more, if they desire. Some other power supplies, such as this 500-watt unit from Clever Power, include a variable speed fan that automatically increases and decreases the spin of the fan, depending on the system's power draw.

Noise/Vibration Isolators

Products are available to reduce the vibration caused by system components as well as to insulate the case to keep noise from escaping. For the bottom of the computer case, rubber feet are available to replace the hard plastic ones generally found. Silicone gaskets can be installed between a power supply or case fan and the case to reduce the transmission of vibrations and the amplification of noise. If you want to keep the noise inside your case, there is even adhesive-backed sound insulation that can be applied to the inside walls of a computer case.

Get tech and computer tips online at Geeks.com.

Jason Kohrs is a technology enthusiast with a background in mechanical engineering. He spends his free time with his wife and daughter or at his online hideout, bigbruin.com.

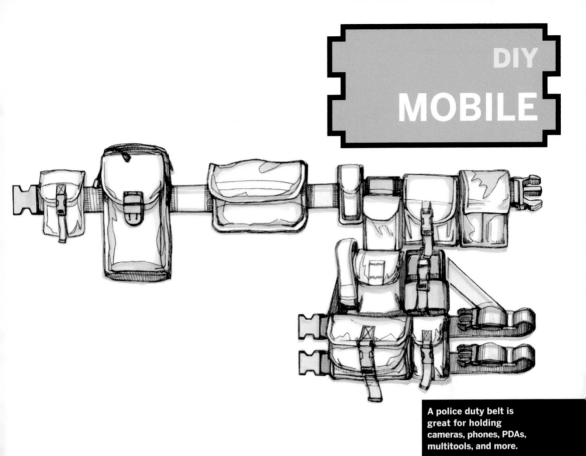

GEARED UP
Use a gunbelt and leg holster to hold your gear.
By Andy Ihnatko

Illustrations by Damien Scogin

We geeks have a crackhead-like dependence on personal electronics, gizmos, tools, and other modern fetish objects. Things like pocket computers, smartphones, LED flashlights, USB thumbdrives, multitools, ZipLinq cables, notepads, digital cameras, spare batteries, and GPS units enhance our lives in obvious and inexplicable ways, but we can't deny that living in the Push-Button World of Tomorrow greatly complicates the otherwise straightforward task of changing one's pants.

Every night, you have to empty all of your pockets. Every morning, you have to fill 'em up again. And portable pockets (in the form of belt pouches) are a mere Band-Aid solution. Unless your electronics are machine-washable, you still need to unthread them from your belt and

reinstall them over and over again, morning after morning. Decent men and women change their pants every day, so what else can you do?

Well, you can head off to your nearest police supply store and buy yourself a genuine, professional duty belt. With your pouches suspended off of that wide band of stiff, thick nylon or leather (structurally speaking, it acts more like a supporting frame than a belt), a simple click of the buckle leaves 6 pounds and $1,200 worth of personal electronics hanging off your bed-post until you get dressed again in the morning. And no, whistling the Clint Eastwood theme from *The Good, the Bad and the Ugly* as you put it on isn't at all inappropriate.

My day-to-day gunbelt configuration consists of a medium-size pouch for my cellphone, iPod,

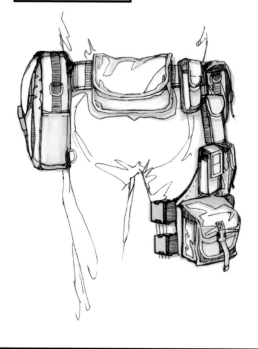

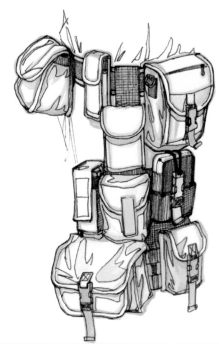

How flexible is a gunbelt system for carrying your stuff? Flexible enough that you'll no longer bristle at a concert event's "No Bags or Backpacks" policy.

I keep a pouch containing iPod speakers, a canister of Pringles, and a thermos of frozen daiquiris ... or as I like to call it, the "Date-In-A-Bag."

and PDA, plus a holder for my Leatherman tool. But with a drawer full of pouches purchased over the years at various camping and photo stores, I can easily add capacity to suit the situation.

For the ultimate in added capacity and convenience, buy yourself a leg holster, which allows you to quick-draw your smartphone, even when you're sitting down or wearing a jacket. Tactical Tailor (tacticaltailor.com) manufactures equipment for urban SWAT units and Army Rangers. They make a "Modular Leg Rig" that can be custom-configured to your specific needs, along with a wide array of pouches that can easily be perverted to nonlethal, geeky needs and will hold everything but your PowerBook.

When I attend trade shows and conferences, my usual gunbelt is supplemented by TT's small leg rig. I've configured it with their adjustable Small Radio Pouch (which is perfect for a PDA or a chunky smartphone), a Small Utility Pouch for my camera, plus the real superstar of their line: the compact, compartmented Multi-Purpose Pouch, flexible enough to hold anything from a folding PDA keyboard to a palmcorder. You can even mount most third-

party belt pouches to the leg rig, using Tactical Tailor's "Malice Clip" system.

Gunbelts are a perfect answer to the blight of personal electronics. I've been wearing one for years, and its value has only increased with recent tightening of airport and building security.

Yes, indeed: I routinely walk through airport security while wearing a police gunbelt and a SWAT tactical leg holster, and I haven't been held in a windowless room without charge even once. To the contrary, screeners and passengers are relieved to encounter a geek who can get all of his personal gear on the conveyor and walk through the archway after just 2 seconds of fiddling with a buckle, instead of holding up the line for 5 minutes while he desperately curses and pats himself down.

Just, um, be sure to refer to your gunbelt as a "utility belt" while you're in the facility.

Andy Ihnatko (andyi.com) is *The Chicago Sun-Times'* technology columnist and the author of a best-selling series of Mac books for Wiley Publishing.

BOOST YOUR SIGNAL
Improving MyFi XM satellite radio reception.
By Bob Scott

Although I love my XM MyFi portable satellite radio, I'm not as smitten with its lipstick-sized portable antenna. It's supposed to provide reception when the internal antenna is shielded from satellite goodness by a backpack, or by your water-filled self. That's the theory, anyway. In my neck of the woods, the portable antenna rarely improves reception.

In contrast, the tiny mag-mount patch antenna for car use pulls in a solid signal despite my dodgy shelf-package installation. It looked like an ideal portable replacement antenna.

My voltmeter provided the good news that the SSMB jack supplies the 4.5-volt bias needed to power the amplifier in the patch antenna.

Fooling around a bit, I found that the car cradle would work without external power, allowing me to do some side-by-side testing of the patch, via the cradle, and the portable antenna in problem portable-reception areas. As expected, the patch not only produced a stronger signal, it also was significantly less fussy about pointing.

Convinced that I now had probable cause to start violating warranties, I cut the cable on a spare patch antenna down to five feet and re-placed the SMB plug with an SSMB. Connecting the hacked antenna to the portable antenna jack on the MyFi, I steeled myself for the worst and switched it on. Three bars on the signal meter!

My new portable patch has worked great in cars, buildings, or attached to my backpack (via a steel washer on the inside of the fabric) with no apparent ill effects on the MyFi. Don't feel like soldering? I just found an equivalent pre-made antenna setup available for about 46 bucks. A little over $15 will get you an SMB/SSMB adapter cable that lets you use the stock patch as a portable antenna, sans the car cradle.

Go on the Go

It's ironic that a security expert is helping people commandeer public PCs to play games. But Karl Sigler, who trains executives in computer security, created a miniCD and USB drive that will override a computer's normal operating system and substitute a tiny version of the Linux system. It sets you up to play Go, the 4,000-year-old Chinese game beloved by programmers and geeks around the United States. "You just pop it in the CD tray and flip the computer on," says Sigler with a certain glee. "I have yet to find a machine it won't boot on."

With a background in building self-contained CDs for computer forensics, Sigler created a Live Linux CD for his favorite new game. The technology allows users to run a whole operating system and software on a computer's RAM rather than a hard drive. Sigler modified a version of the operating system called DamnSmallLinux, added dozens of Go applications in Linux, and recompiled it back to a CD so that it's executable. Using the disc, absolute beginners can play against the machine or get an interactive lesson; fanatics can enter online games or pull up descriptions of 4,000 historical games to hone strategy. And you can run the disc in most locked-down PCs in cafes, training stations, and libraries. "When you pull out the CD, the computer works normally and there are no fingerprints that you were there," says Sigler.

—*Bob Parks*

>> **Live Linux CD: hikarunix.org**

Photograph by Anne Sigler

Bob Scott is a statistical construct of various consumer electronics marketing departments.

WEIRD USB
Plug into a world of novelty.
By Phillip Torrone

USB is the "cable access" of modern computing in more than just a literal sense. Here's a look at a variety of USB gizmos, from the outlandish to the outright useful.

Ghost Detector: You know Scooby and gang need more than just to store photos of the mean old men who run the amusement parks on some type of USB drive — this one detects ghosts, too. Zoinks! makezine.com/go/ghostUSB

Ashtray: For that smoker who has everything, a USB ashtray. We're waiting for a version that emails your insurance company. makezine.com/go/ashtrayUSB

Self-Destruction Box: When the feds bust in looking for that Bit Torrent collection you've been squirreling away, the USB destruction box is the way to go — quickly and discreetly hit the nuclear option. makezine.com/go/selfdestructUSB

Light-Up Hub: What's the sense of transferring all this data unless you can see it? Routers get the blinky; why not a USB hub? makezine.com/go/lighthubUSB

Light-Up USB Cables: For the modder in all of us. That new blinged-out PC is drab with IBM-grey cables. These light up! makezine.com/go/lightupUSB

Laptop Cooler: If you're male and own an Apple PowerBook, you need this. Enough said. makezine.com/go/coolerUSB

Hot Plate: Why go all the way to the kitchen to keep that Hot Pocket warm and toasty? With the awesome 5V cooking action of USB, you can stay parked at the desk. makezine.com/go/hotplateUSB

USB for Rock Stars: USB fobs are handy, but not something you wear rocking out Bo Bice style. Now you can. makezine.com/go/rockUSB

Heating Pad: Nothing cramps up your back like staring at an RSS reader all day, but the sweet warmth of the USB port makes it go away — at least until the next refresh. makezine.com/go/padUSB

Disco Ball: Get on down with the USB badness, if you can dig it. Also makes the cat act weird. makezine.com/go/discoUSB

Eye Massager: Eyes hurt from staring at the screen all day, but don't want to leave the umbilical cord? We see a USB Eye Massager in your future. makezine.com/go/massageUSB

Battery Recharger: Why in the world is this available only in Japan? We'd actually use this. makezine.com/go/chargeUSB

Phillip Torrone is associate editor of MAKE.

THE WORLD AS YOUR CANVAS

Use GPS to create giant-sized works of art.
By Julie Polito

Sometimes you need something bigger than a piece of paper to express your artistic self. Like, the whole world. A GPS device, a little planning, and a lot of free time can help you leave a giant footprint — or any other shape — on the planet. Think of it as crop circles for the 21st century.

Aaron Roller of Sausalito, Calif., wanted to give his girlfriend something extra special for Valentine's Day. He'd been toiling at a start-up company all year and she'd been an extra good sport. Roller is the chief technology officer for Motion Based, Inc., a company that created a web-based application that turns GPS data into functional analysis and online mapping for train-ing athletes. He realized that those patterns on the map could also translate into pictures on a giant Valentine card.

After a little visualization, Roller set off into San Francisco Bay with a twitchy river kayak, a GPS, and a dream. "I don't think I'd been out on the bay more than a couple of times in this kayak, but I had a mission," he says. Half an hour and two miles of paddling later, Roller had traveled the shape of a large heart and uploaded the image of his path onto the Motion Based map-

Tic-Tac-Toe: A GPS unit and a mapping app can turn the world into a giant Etch A Sketch.

The GPS unit on your car's dashboard serves as a "stylus" as you drive your car around the urban canvas. You can also use a bike or take to the water in a boat.

ping site (motionbased.com). "I had to be careful not to turn too sharply," he says. "And there is a little crook toward one end of the path where I had to navigate around a big boat in my way." His girlfriend loved the heart and the magnitude of the message.

While Roller says he is a fan of GPS drawing, he considers himself a dilettante. He points to GPSDrawing.com as an example of how elaborate you can get with your ideas. The site has several examples of impressive GPS art, including a large dollar sign in the Las Vegas desert and a larger-than-life tic-tac-toe game played on the streets of Hollywood (an exercise that took four hours of driving over a 23.6-square-mile pattern). Jeremy Wood, one of the founders of GPSdrawing.com, mapped out the grid the night before with his brother, and the two of them played the giant tic-tac-toe and mapped out their moves as they drove.

Wood has been drawing with his GPS unit since the day he bought it, and says he is merely building on an idea that is as old as humankind. "There is evidence of the concept all around us," says Wood. "Traces of trajectories can be found throughout history. They grow from footpaths to roads when we move."

If you're inspired and planning to create a little GPS art of your own, here are a few pointers to get you started.

First, grab a GPS device. Any one will do, really, as long as it maps your every move and uploads to a software program. Aaron Roller used a Garmin eTrex Vista C GPS, which costs around $428 and gives you a 256-color readable map display to keep you on track during your journey. You can also opt for a more portable GPS like the Garmin Foretrex 201 if you're traveling light. The Foretrex 201 costs around $150 and fits sportily on your wrist, but lacks the easy navigational screen of the eTrex Vista C.

Design your image, and keep it simple. Save your detailed likeness of the Virgin Mary for your sketchpad or grilled cheese sandwich. The more complex the design, the more time and editing you'll need to execute it.

For your canvas, choose an open space with very few obstacles. If you're planning a large route by car, remember that you will be restricted to usable roads. Walking in a field or paddling on an open body of water will give you flexibility.

Map out your route with a mapping program. Roller recommends tools such as Expert GPS, MapSource, or National Geographic Topo to help you plan your path. "You can download the route into your GPS and just follow the arrows," he says.

When you're following your route, think of an Etch A Sketch. "Start at a point, and draw a single continuous line," Roller recommends. You may need to backtrack to stay in your pattern.

After you've finished your design, upload your data into a software mapping program such as Motion Based (motionbased.com). Jeremy Wood uses a custom-designed program called the GPSograph, designed by fellow GPS artist Hugh Pryor. It's available on the GPSdrawing.com site and enables 3D rendering and map/aerial image alignment. Motion Based has editing tools you can use to clean up the drawing and erase start and finish points and any glitches.

Julie Polito is a writer whose pieces have appeared in *Salon* as well as *Self*, *Mobile*, and *San Francisco* magazines.

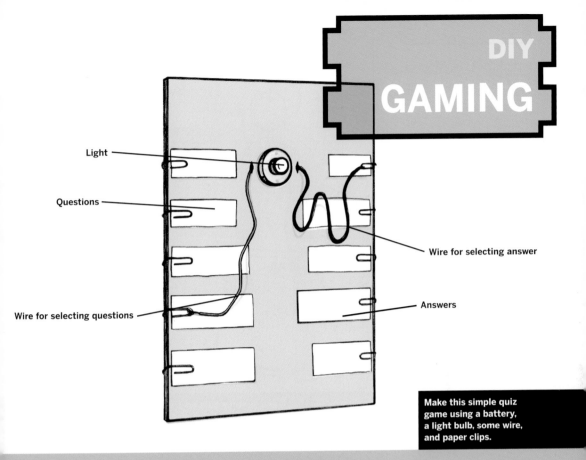

Light

Questions

Wire for selecting answer

Wire for selecting questions

Answers

Make this simple quiz game using a battery, a light bulb, some wire, and paper clips.

CIRCUIT QUIZ GAME
Teach kids about circuitry in 30 minutes.
By Ben Wheeler

Arthur C. Clarke's Third Law states that "any sufficiently advanced technology is indistinguishable from magic." Clarke's law could use a corollary: sometimes it doesn't take much for something to be advanced enough to seem mystical. Today's kids are used to every teddy bear and softball being Turing-complete, so hidden complexity doesn't impress them. To really dazzle them, you need older technologies, like circuitry, which feel magical by their very openness and simplicity.

Here's a classroom-tested example: a simple project that introduces kids to circuitry. It functions as a quiz game, where a light bulb signals that the correct answer has been selected. The circuit itself is very basic, but if you provide enough sections of quiz material, it's surprising

and satisfying to see it work. You can quickly rearrange the questions and swap them in and out, which makes for fast competition, impromptu topic-creation, and ample opportunities for ridiculing siblings.

Questions and answers are attached to a board by paper clips, with matching question-answer pairs connected in back. Two wires that dangle on the front of the board connect to the light bulb and the battery. When a player touches one of the wires to a question and the other to its correct answer, the circuit completes across the hidden wire behind the board, and the bulb lights.

Circuitry projects like this are also a great way to introduce kids to circuit diagrams, and show how they map to physical reality. Here's a simple circuit diagram for this project:

Illustrations by Damien Scogin

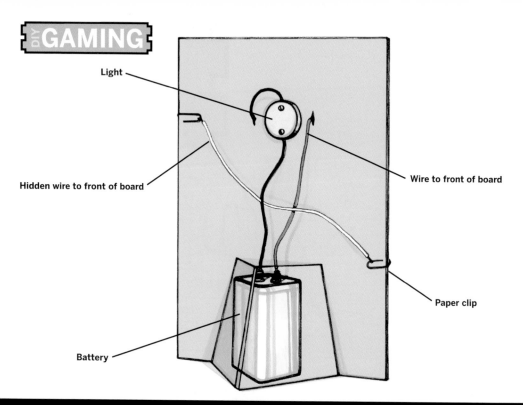

Light

Hidden wire to front of board

Wire to front of board

Paper clip

Battery

Behind the curtain: Compare the illustration on the previous page with the one shown here. The two wires on the front of the board (emanating from holes on either side of the light bulb) form a complete circuit and light the bulb only when they touch the paper clips that are connected by the wire behind the board. (For purposes of clarity, the paper clips for the wrong answers are not shown in this view.)

All of the circuit components in this project are available at RadioShack.

MATERIALS

6V light bulb (RadioShack #272-1128)
Light bulb base with screw terminals (RadioShack #272-357)
6V lantern battery (RadioShack #23-016)
Wire spool: 20-gauge, insulated (RadioShack #278-1225)

Two large pieces of cardboard
Paper clips
Duct tape
Paper and pens or printer, for labels

Total cost: $15

TOOLS

Hobby knife
Small screwdriver
Wire cutters/stripper

Make the Board

Cut two identical cardboard trapezoids, about 6" on the top and 9" on the bottom, with sloping sides. Tape them together along one side and to the bottom of the quiz board to form a stand.

Install the Light

Cut a small circle in the center of the quiz board for the light bulb base. Cut two small slits on either side of the circle. Attach two wires to the light base. Send one wire through a slit to the front of the quiz board. Screw in a bulb and fit the base into the circular hole.

Install the Battery

Put the battery inside the quiz stand. Attach the second wire from the bulb to a battery terminal. Connect a long wire from the other terminal through the other slit to the front of the board.

Wire the Answers

Make each question-answer wire by wrapping both ends of a wire around paper clips. Make sure the connections are snug. Attach the clips on alternate sides of the back of the quiz board.

Label the Quiz

Make paper labels for the quiz questions and answers and place them next to the corresponding paper clips. Place dummy paper clips next to all the non-matching questions and answers, so they all look identical from the front.

Ben Wheeler is a math teacher living in Brooklyn, N.Y.

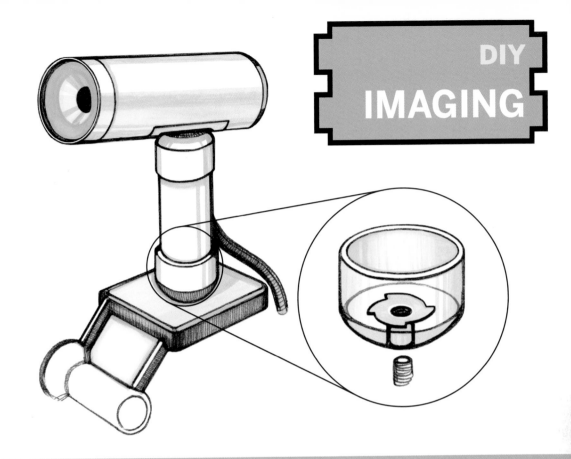

Illustration by Damien Scogin

iSIGHT TRIPOD MOUNT
Point your webcam anywhere you want.
By Steve Vigneau

In mid-2003, Apple announced a FireWire camera called the iSight. With true autofocus and the ability to capture 640x480 video at 30 frames per second, these cameras have become quite popular with Apple users. However, one of the camera's most innovative features — the mounting hardware — is also its most irritating quirk.

The iSight uses a bracket that supports one end of an Apple FireWire cable, and this cable snaps into a FireWire port on the bottom of the camera. Because the iSight is lightweight, this configuration works well in certain instances. Apple ships a number of attractive brackets designed to mount the iSight on a CRT, laptop, or flat panel monitor. What these brackets do not allow for is mounting the iSight on a tripod.

As I often want to point the iSight out a window

or someplace other than at what is sitting in front of my monitor, I needed a way to mount my iSight on a tripod. There seems to be a lack of commercially available mounting options (Kaidan makes an adapter for $20, but it requires an Apple-branded FireWire cable and part of an iSight bracket), so I headed to the basement to build my own. Here's how you can do the same:

Beyond some basic tools, you will need about three inches of ¾" PVC pipe, two end caps to fit the pipe, a FireWire cable, a ¼-20 tee nut, hot glue, and some PVC cement.

First, drill one of the PVC end caps with five holes so that the tee nut and its wings will fit inside, points down.

A tripod bracket for your iSight gives you the freedom to place it anywhere you want.

Fig. 1

Fig. 3

Fig. 2

Fig. 4

Making the iSight tripod mount, step-by-step: Drill holes in one of the PVC end caps. The four smaller holes are for securing the tee nut, while the larger hole is for the tripod's mounting screw (Fig. 1). Squirt hot glue into the end cap to secure the tee nut (Fig. 2). Cut a rectangular hole in the other end cap, making sure the FireWire connector protrudes enough for the iSight to plug all the way in (Fig. 3). Make a notch for the cable (Fig. 4).

I used a ⁵⁄₁₆" bit for the center hole, and a ³⁄₃₂" bit for the four surrounding holes. Press the tee nut into the holes inside the end cap and secure with hot glue.

Cut a rectangular hole in the center of the other end cap, through which the cable's FireWire connector can fit snugly.

Next, cut a slot up one side of the pipe. This slot should be as wide as the cable, and as long as an end cap is deep, plus the thickness of the cable. Using PVC cement, fasten the end cap with the FireWire connector hole to the opposite end of the pipe from the slot, aligning the rectangular hole with the slot.

Orient the keyed side of the FireWire connector away from the slot in the pipe and fit it up through the pipe and into the rectangular hole in the end cap. Allow it to stick out a bit so the iSight will have room to tilt freely. Tack the FireWire connector in place, inside the tube, with a bit of hot glue.

Bend the cable so it runs out the slot in the side of the tube, and fill the body of the tube with hot glue. This secures the connector and provides strain relief for the cable.

Quick Tip: USB Cam and iSight

Apple's iChat AV application requires a FireWire camera, but if you already own a USB webcam (which can be purchased for as little as $15 online) and don't feel like shelling out $130 for an iSight, you can buy a utility to fool iChat into thinking you're using a FireWire camera.

Called iChatUSBCam (ecamm.com), this $10 program allowed my sister in Colorado to use her el-cheapo Logitech webcam to videochat with me for a fraction of the cost of a new iSight.
—Mark Frauenfelder

After the glue has cooled, cement the end cap containing the tee nut onto the other end of the pipe, and you're done. Screw your tripod's connector into the bottom of your new iSight Tripod Adapter and enjoy the myriad of places where your iSight can now be pointed.

Steve Vigneau makes, bakes, and un-breaks things around Southeast Michigan.

Photograph by Steve Vigneau

The Fauxlance Photographer

How to get VIP treatment by dressing the part of a pro photographer. By Andy Ihnatko

Photograph by Andy Ihnatko

Being a columnist for a Great Metropolitan Newspaper has its perks, first among them being the fact that the *Sun-Times* and I have worked out a little agreement by which they pay me money for my columns.

But there's another benefit. My idea of a fun time is to walk around with an SLR and make some pictures. Sometimes, getting Just The Right Shot means climbing up on a balcony or asking a street performer to turn toward the sunlight or standing rock-still with my SLR against my face, waiting for the scene to finish composing itself. And being able to flash a business card or a recent clipping is often the difference between being told "I'm sorry, but I'm really going to have to ask you to leave" and being told "You'll get much better shots of the parade from the top

of the reviewing stand ... just follow me."

If you're not a columnist, well, simply dressing the part of a working journalist can go a long way. The way you appear and the way you behave subconsciously communicate to people that you're not some creep with a camera or just a tourist getting in the way. It's not about misrepresenting yourself; you're just subtly encouraging people to treat you with the same amount of respect that you invest in your photography.

Here's how you can get treated like a pro:

Overall Dress

Don't look like a bum, unless you actually are a photographer for *Sports Illustrated*.

As a freelance journo, you don't have the money for expensive clothes and they'd just get ruined as you travel from event to event anyway. So the effect to aim for is "as comfortable as you can get away with while still looking professional and presentable." Meaning: buttoned work shirt (no tees or pullovers), casual slacks, and sneakers that can pass for shoes.

Camera (general)

Yup, it's bigotry, but civilians instantly respect the Big Black SLR. A line of servers at a Chowderfest will not patiently pose for you if you're packing an $80 point-and-shoot.

Pros almost always wear two cameras (one with a short lens, one with a tele). Finally, a reason to get your old film Nikon out of mothballs! Always wear them across the shoulder on a long and comfortable strap.

Camera Lens

Put a lens shade on your camera, like every working photographer — not to combat sun flares, but to protect the glass and filter threads against bumps.

Camera Flash

An external flash is a must, even when shooting outdoors. For extra credibility, rubber-band a white index card around the head as shown. It's a cheap bounce-diffuser for better lighting.

Bag and Waistpack

Photographers need to get their hands on another lens, memory card, or filter without any fiddling around. So: A waistpack is essential, and if you wear a bag, it shouldn't be a backpack.

Necktie (in bag)

Handy to have, in case you discover that most of the other media are wearing 'em. A rare problem, but I've gotten into at least one event because I was able to put a tie on.

Microphone (in bag)

If you're recording audio for a podcast, get a mic that is, or looks, impressive. And, the shabbier the windscreen, the better.

Old Press Pass (in bag)

OK, it's a cheap trick, but I might have a press pass in my bag, too. If I'm talking to the media coordinator of an event and I can sense that

they're on the fence about giving me a pass, I'll rummage through my bag for something and whoops, a genuine laminated badge to a previous event fell out! Sorry about that. Hmm? Oh, yes, that's from a space launch that I covered, thanks.

More important than a pass is a batch of business cards. No need to lie: just use your name and put the simple word "Photographer" underneath. Include your address, phone number, and email, and a link to your photo site. Photo subjects and media coordinators want to be reassured that you're for real and not some creep; 500 cards for $10 goes a very, very long way.

Breast Pocket

Keep it filled with random crap at all times.

Notepad

Hands-down the most important prop in the picture. All working media need to have a notepad handy at all times. Not a PDA — a notepad. You're always scribbling down names of people you've spoken to or photographed and millions of other notes, and it also serves as documentation that you actually were where you said you were, in case lawyers get involved later on.

Plastic Bag

If you're still shooting film, you gotta keep it in Ziploc baggies. It's just the rule.

Important Notes:

So what about getting an actual press pass? Well, it can't hurt to ask (call the event office and ask for Media Relations). Two big warnings, though: First, do not, do not, do not lie. If I'm not attending an event with the express purpose of covering it for my paper, I'm crystal-clear about my intentions. Trust should never be abused. And of course you want to get into the media tent, with its air conditioning and water and hookups. But stay out. For members of the actual media, it's essential. If I have 30 minutes to file my story but you're using the last network connection to play some MMORPG ... well, look, it'll get ugly. Finally, give away your photos and story to your local paper. Remember there's just one difference between an amateur photographer and a pro: a single published clipping.

Andy Ihnatko (andyi.com) is the *Chicago Sun-Times'* technology columnist and the author of a best-selling series of Mac books for Wiley Publishing.

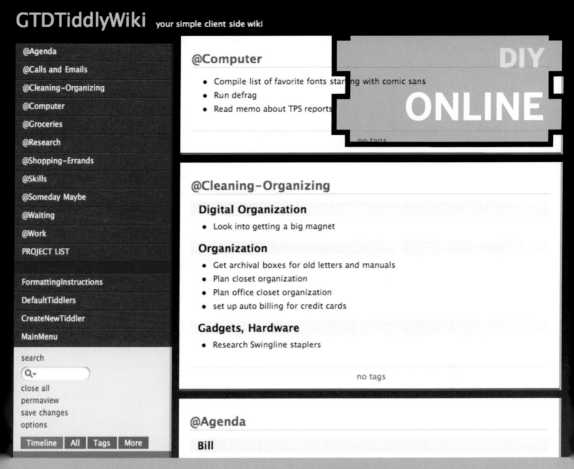

WIKI YOUR WORLD

Getting things done with GTD TiddlyWiki.
By Gareth Branwyn

It's a sad commentary on hardware and software design that few products deliver on their promises to make a meaningful difference in how we live and work. When something does rise above the clutter of flashy, but ultimately distracting, widgets and truly changes and inspires you, it's a genuine thrill. When such a discovery is free, the thrill is even headier. Such is the case with Nathan Bowers' browser-based, cross-platform application GTD TiddlyWiki.

Wikis Writ Small

You've probably noticed that wikis, open source collaborative web publishing tools, have been gobbling up a lot of mindshare recently. High-profile projects such as Wikipedia are bringing the wonders of wiki to the world. And the collaborative,

open nature of wikis has spawned numerous mutations on the basic theme of quick ("wiki-wiki" is Hawaiian slang for "quick") web publishing.

One of those variants is TiddlyWiki, a client-side app by Jeremy Ruston that exists as a single HTML document on your web browser. Users can create a series of hyperlinked Post-it Note-like entries, called Tiddlers, using an extremely simplified wiki markup language. One of the most popular TiddlyWikis is Bowers' GTD TiddlyWiki.

GTD refers to *Getting Things Done*, David Allen's best-selling, geek-friendly book and personal task management system. Its appeal to a lot of geeky GTD users is in breaking down and tackling (or deferring)

Use this personal wiki to manage your tasks and plan your strategies.

actions as discreet components, and using a reliable storage system to hold and retrieve these "packetized" tasks. This fits in perfectly with the heart of TiddlyWiki, the Tiddler.

TiddlyWiki started out life as an experiment in hyperlinked microcontent. In the early days of hypermedia, there was Bill Atkinson's breakthough HyperCard, Apple's multimedia-capable, hyperlinkable application. HyperCards were discreet, non-linear ... well, cards. The web supplanted HyperCard and offered a globally hyperlinked universe of media, and the microcontent of HyperCard went macro. Then blogging came along and laid a dated journal/log book metaphor over the web. Both the web in general and blogs in specific are hyperdocuments, but they tend to be used in a rather linear manner. The idea of TiddlyWiki is to reintroduce the small, hyperlinked "card" metaphor. And, as many people have learned, it turns out to be ideal for managing tasks.

Getting Things Done with Your Wiki

To set up a GTD TiddlyWiki (or GTDTW for short) on your desktop, all you have to do is go to the app's website (shared.snapgrid.com/gtd_tiddlywiki.html) and save a copy of the HTML page to your hard drive. The entire contents of your TiddlyWiki — oodles of JavaScript and Cascading Style Sheet code, along with your content — is saved as this single document.

GTDTW comes pre-wired with a series of categories based on the system described in Allen's book, but you can easily tweak away on your own. It takes only a few minutes to set up the app, learn the simplified wiki formatting (a built-in Tiddler teaches by example), and customize the interface and content to your tastes. Once you get the hang of it, you'll be hyperlinking Tiddlers like crazy, launching apps from your wiki, hyperlinking to web content, pasting images into Tiddlers, making sortable tables, and all sorts of other "sophisticated" functions.

The power and simplicity of GTDTW are darn-near awe inspiring. It looks sharp, runs with a responsive snap you don't find in most applications, and is really fun and easy to use. It does what it's supposed to do, reliably, and it's got a rapidly growing user community that's dreaming up new hacks on a seemingly daily basis.

Getting Even More Done

If you want to really customize GTDTW, you'll need to roll up your sleeves. GTDTW is client-side only, so unlike other wiki apps, it is not collaborative. There are server-side TiddlyWikis that you probably could set up as a collaborative GTD tool (such as TiddlyWiki Remote, phiffer.org/tiddly).

Another problem, more of an annoyance really, concerns the printer function. GTDTW lets you print Tiddlers onto 3x5 cards, creating literal, analog microcontent cards. This allows GTDTW to integrate perfectly with the HipsterPDA (hipsterpda.com), another popular and refreshingly retro implementation of GTD concepts. But on browsers where you can't easily turn off headers and footers (such as Firefox on the Mac), your lovely Tiddler cards have unsightly text on the top and bottom. This is fixable via various hacks. For workarounds on this and other GTDTW bugs, and to get the latest hacks and variants, check out the forums at groups-beta.google.com/group/GTD-TiddlyWiki.

After reading *Getting Things Done*, I tried a number of GTD apps and hacks for using existing personal management tools (such as Gmail). Nothing (except the HipsterPDA) stuck, and they all seemed to get in the way of, well, getting things done. After several months of using GTDTW, I'm amazed at how much it's improved my organizational life and my day-to-day productivity. It's also a delight to use, and the whole microcontent concept underlying it has inspired a possible book idea. When's the last time Entourage did that?

Note: The GTD TiddlyWiki I actually use is one called GTD TiddlyWiki – KnipSter Enhanced (knipster.net/projects/gtd_tiddlywiki.html). The main enhancements are sortable grid tables and a dated journal entry feature.

Gareth Branwyn writes about the intersection of technology and culture for *Wired* and other publications, and is a member of MAKE's Advisory Board. He is also "cyborg-in-chief" of Streettech.com.

TEN DOLLAR PSEUDOSCOPE

See everything inside out through this classic optical instrument. By Rob Hartmann

Photograph courtesy of Rob Hartmann

The pseudoscope is a device that plays a trick on the eyes, switching the perception of near and far by reversing stereoscopic vision. It was invented by the great Victorian-era scientist Charles Wheatstone, and M.C. Escher used one to help create some of his famous perspective-bending illustrations.

You can buy one of these unusual devices for about $800 from Grand Illusions, so I decided to build a simple one of my own. Actually, I've built a couple of them now, with varying degrees of complexity, but here I'll show you how to make the easiest one for much less money: about $10.

Assemble It

I built the pseudoscope the way you'd guess just by looking at it. First, I measured and drilled the holes in the board. Then I drilled the cubes, glued the mirrors to the cubes, and screwed the cubes onto the board. I didn't glue the cubes to the baseboard because you need to be able to rotate the mirrors to align them depending on the distance to the subject matter.

Countersink the holes in the base so that the screws won't scratch table surfaces. Also, drill the holes in the cubes before you attach the mirrors.

Wooden cubes from craft stores, like the ones I used, aren't all perfectly square. For each cube,

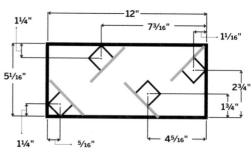

Fig. 2

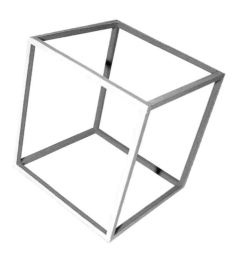

Fig. 1

MATERIALS

Two 3"x3" mirrors
Two 3"x4" mirrors
Four 1¼" wood cubes
One piece of wood,
5"x12"x¼" thick (make
sure it's perfectly flat)

Four ¾" flathead screws
Glue

TOOLS:

Drill and drill bits
Screwdriver

M.C. Escher did it: With four mirrors mounted at exact right angles, you can easily create powerful illusions. Here are the dimensions I used for the screw holes (see Fig. 2). I don't think it's critical to use these exact dimensions; being a fraction of an inch off here or there shouldn't matter much.

you need to find two adjacent faces at exact right angles from each other, one side to serve as the base of the cube, where you drill the hole, and the other side to glue the mirror onto.

To find the good sides, check with a right triangle or push cubes together on a flat surface at different orientations until you find two that sit perfectly flush against each other. Just hope that the two cubes aren't off-angle by perfectly matching amounts.

To keep the mirrors at exact right angles, use a thin layer of glue, and glue them in pairs, pressing the blocks and mirrors together with the mirrors face-to-face.

Experiments and Enhancements

Your brain uses multiple cues for depth perception rather than relying entirely on comparing left eye versus right eye. As a result, inverted perceptions from looking through a pseudoscope can appear and disappear, depending on what you're looking at. The effect generally works better when you're sitting still rather than moving, and it's fun to experiment with what works and what does not.

One vivid pseudoscope experiment is to watch framework solids rotate, especially if their inside and outside faces are of different colors. I built a framework cube using twelve 6"x¼"x¼" pieces of basswood, and painted the inside faces red and the outside faces white (see Fig. 1). I hung the cube up and slowly spun it; through the pseudo-scope, the cube looked red on the outside and like it was spinning in the opposite direction. When I closed one eye, the cube reverted back to its normal appearance. This illusion is best seen with strong lighting.

Always clean the mirrors before using the scope — fingerprints and smudges will greatly diminish the illusions. To take things up a notch, you can use front-surface mirrors, which eliminate the slight ghosting effect of conventional mirrors. Prices for these vary widely, so you should query different glass shops and look online.

I found some slightly larger front-surface mir-rors for a good price at C and H Sales (aaaim. com/CandH), and used them to build a fancier pseudoscope with a custom-shaped walnut base.

Rob Hartmann is an electrical designer who lives in Fairfield, Ohio and enjoys mechanical puzzles and science projects.

The Davis Vantage Pro2 collects weather data and sends it back to the console for web upload.

CITIZEN WEATHER STATION
Collect meteorological data on your roof and donate it to science. By Terrie Miller

I love the idea of citizen science, where amateur enthusiasts collect and combine data in the spirit of shared scientific discovery. After learning about citizen weather observer projects, I got the itch to plant my own weather station and start contributing data. I decided to go with a commercially manufactured station, and a local group of weather geeks recommended the Vantage Pro2 from Davis.

The Vantage Pro2 measures precipitation, temperature, humidity, wind speed, and wind direction (and you can add additional sensors to measure solar radiation and leaf and soil moisture). There are two pieces: the station itself, which sits outside and collects the data, and the console/receiver, which hooks up to your computer via USB. The station is solar powered and transmits weather data to the console/receiver wirelessly, so you

don't need to worry about getting AC power cords or data cables up to your roof. Davis' WeatherLink software, sold separately, downloads data from the console, and lets you upload it to your own website and contribute it to the Citizen Weather Observer Project (CWOP), a grassroots, private-public partnership that collects and centralizes meteorological data from stations like mine.

It was easy to assemble; it took about an hour to put together and set up the communication to the console. Depending on your circumstances, the optional mounting tripod might make installation easier. We installed the weather station at the O'Reilly offices, but we had to attach it to an outside wall instead of putting it directly on the roof. We needed some extra height in order to get the station up into unobstructed wind, so we got

The Davis Vantage Pro2 Weather Station includes the sensor unit you install outside (above, left), and a **console unit that you keep indoors (above, right). The sensor unit is easy to assemble.**

a 10-foot conduit pipe from the hardware store, attached that to the wall, and then mounted the station at the top. When you install the station, you'll need to know which direction is north, so you can align the anemometer to correctly record the wind direction.

The console/receiver lets you view current conditions recorded by the station, as well as some past conditions. You configure the console with your latitude, longitude, altitude, and other settings that relate to the station itself. The console also displays current sunrise and sunset times and a rudimentary weather forecast. I liked that it displays the rainfall rate as well as the cumulative rainfall, and that during a downpour the console reports, "It's raining cats and dogs!"

According to the station's specifications, it can transmit up to 1,000 feet to the console line-of-sight, or 200 to 400 feet through walls. We were concerned about our installation at O'Reilly, where the building's steel frame interferes with wireless, but we had no problem transmitting data from the roof high above the third floor down to the second floor.

The Weatherlink software makes uploading

data to the web or CWOP quite easy (it supports FTP uploads, but sadly not SFTP). The web template system generates a number of interesting charts and graphs from your data. The default HTML templates could use a designer's touch, but I found them easy to customize.

When I signed up for CWOP, our station got its own call letters: CW3724. Now anyone can access its data, and it's interesting to compare our station's readings to those of other citizen stations nearby. Seeing a storm roll in has never been so fun!

Davis Wireless Vantage Pro2 Weather Station, $595, davisnet.com/weather/products/ weather_product.asp?pnum=6152
Current data from our station:
makezine.com/weatherlink
Current station data, reported through CWOP:
findu.com/cgi-bin/wxpage.cgi?cw3724
NOAA Citizen Weather Observer Project:
wxqa.com

Terrie Miller works at O'Reilly Media, Inc., and publishes PointReyes.net and CitizenSci.com.

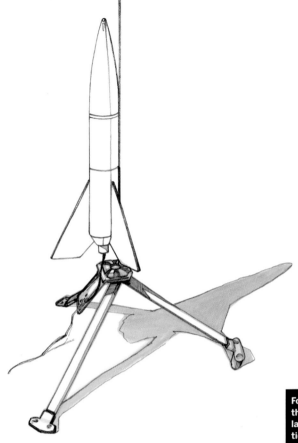

Illustration by Damien Scogin

LAUNCHING LIGHT
Portable, collapsible model rocket launch pad.
By Stefan Jones

My new dog is terrified of fireworks. This wouldn't be a problem outside of July 4th — if I didn't fly model rockets. But she can't tell the difference between an Estes B6-4 and a Joyous Family Whistling Fountain, so rockets terrify her too. This makes weekend trips to the organized launches in central Oregon impractical, so I've been flying locally. Launching in suburbia can involve hikes to unused fields, so I designed this sturdy, portable launch stand made of PVC pipes and fittings.

Preparation
Legs: Cut the PVC tube into 1' lengths. Sand the ends smooth.

Feet: Drill a ¼" hole through the side of each T-junction.

Hubs: The corner junctions form the hub of the launcher's tripod and support the launch rod. Stuff the center of each corner junction with epoxy putty, taking care not to get any in the socket areas. The hardened putty provides additional weight and support for the launch rod.

After the epoxy has cured, drill a hole straight

MATERIALS
⅞" non-plumbing-grade PVC pipe, 3' long
⅞" PVC T-junctions (3)
⅞" PVC corner junctions (2)
Epoxy putty
Galvanized metal shingle, 5x8"
⅛" steel rod, 3' long
³⁄₁₆" steel rod, 3' long
Skewer-type tent stakes (3)

TOOLS
Hacksaw
Vise-grip pliers
Sandpaper
Hand drill or drill press (preferred)
⅛", ³⁄₁₆", and ¼" drill bits

Drill a hole down the center of the corner junction's rounded "elbow" for different launch rods. At the launch

site, thread a tent stake through the holes in the T-junctions and press into the ground.

down into the center of the junction's rounded "elbow" and into the putty, taking care not to punch through the other side. Drill different-sized holes in each hub for different launch rods, one ⅛" and the other ³⁄₁₆".

Blast Deflector: Fold over the ends of the metal shingle and crimp flat. Drill a ¼" hole about 2" from one end.

Assembly

You can string the pad's legs and fittings into a tube that is easy to carry, stows in a car trunk, and doubles as storage for your launch rods. Toss the blast deflector, tent stakes, and spare hub can in a tote bag, along with your launch panel, motors, and other gear.

When you've reached the launch area, take the tube apart and assemble your tripod:

1. Choose a center hub and insert the legs into it.
2. Put a T-junction on each leg, turned so a side is flush against the ground.
3. Thread a tent stake through the holes in the feet and press into the ground.
4. Insert one end of the launch rod through the hole in the blast deflector and into the hub.

To angle the pad, pull up a stake and place an object under one of the feet.

Wherever and whatever you fly, be considerate and follow the NAR safety code (nar.org/NARmrsc.html) that keeps the hobby legal!

Stefan Jones dabbles in net-futurism, model rocketry, and gaming.

Don't DIY

Ethan Zuckerman isn't afraid to admit he's wrong. One recent project posted on his blog, a wood-fired hot tub built almost solely from materials he had lying around, fizzled dramatically. Photos show his partner-in-crime, Nate, sitting in the smoky, lukewarm hot tub during a snowstorm, looking stoic at best. Multiple reader responses point out just how dangerous such a project is, prompting Ethan to respond: "Let me be very explicit: **Don't do this.**"
makezine.com/go/hottub

Insert firefly in plastic bubble and a $5 photodiode chip converts light to frequency.

Photograph by Kirk von Rohr

FIREFLY METER

Bioluminescence detector lights way toward insect-cyborg pollution sensor. By Christopher Holt

For me, the memory of catching fireflies in a jar brings me back to the dog days of my childhood summers. Now I'm a part-time biology graduate student, and one of my research projects focuses on the impact of air pollution on firefly flash duration, intensity, and period. Understanding this link could enable us to use fireflies as environmental contaminant sensors, providing an indirect method of monitoring air quality.

If you couple the fireflies with a low-cost way of measuring their bioluminescent properties, you have a sensor. Fusing whole organisms with hardware in this fashion is an exciting new research area, which has produced sensors capable of measuring biological and chemical agents at part-per-trillion concentrations.

Fireflies, or lightning bugs as they are commonly called, are technically neither flies nor bugs, but beetles of the order *Coleoptera*. There are more than 2,000 species worldwide, with 170 species in the United States alone. Fireflies use their light to find mates, which makes them among the few nocturnal insect types that discriminate mates visually. Numerous other insects are luminescent, but fireflies have the rare ability to turn their bioluminescence on and off. Typically, the males will signal with a specific flash pattern, and then wait for a female to respond with another flash pattern.

Most instruments for measuring bioluminescence are bulky, require advanced signal conditioning circuitry, and are far too expensive for a self-funded grad student. I needed an easy-to-use instrument that could measure firefly flash duration, intensity, and period. I surveyed my electronics workbench (yes, I'm probably the only biology grad student with one of these), and my eyes wandered to a Basic Stamp 2 (BS2) Board of Education (parallax.com). I pondered the possibilities of using this simple, easy-to-use microcontroller kit for my bioluminescence meter.

On a few robotics projects in the past, I had already interfaced cadmium sulfide (CdS) photoresistors with the BS2 to measure light intensity. So first I thought I would just plug a photoresistor into a breadboard, build a chamber to house the firefly, and collect data until the cows came calling.

I soon discovered a big problem with my CdS photoresistors: long response and recovery times. In some cases, it took the photoresistor up to half a second to recover its value after an LED flash.

I figured out that Texas Advanced Optoelectronic Solutions (TAOS, www.taosinc.com) had basically solved my problem with their $5 TSL230 light-to-frequency converter. This 8-pin chip has a photodiode and built-in signal conditioning circuitry that does all of the analog-to-digital conversion on board. Also, the chip can be programmed to measure light for a specified duration, and transmit serially to a microcontroller or PC. With this key piece, I had what I needed.

Let's Build It

The hardware components needed for making the firefly flash meter are simple: a BS2 microcontroller, 0.1µF capacitor (to reduce noise), the TSL230 light-to-frequency converter chip, firefly test chamber, and a computer.

For software, I used StampDAQ (Stamp data acquisition) to translate the data coming via serial cable from the BASIC Stamp microcontroller into a format that's usable by Microsoft Excel on the computer. StampDAQ software is available as a free download from parallax.com.

For my delicate measurements, I set the TSL230 pins S0, S1, S2, and S3 high (+5V), which tells the chip to run at maximum sensitivity and resolution. I assembled the circuitry right on the Board of Education, which is an easy way to prototype.

I built the firefly test chamber out of a plastic pipette (VWR #14670-149). I found that cutting the bulb of the pipette approximately ½" from the top provided adequate space for the firefly.

To make sure the firefly had adequate air supply,

```
Code to interface the BS2 to StampDAQ:
'{$STAMP BS2}
'Define variabless

light   var   word          'stores light intensity
sPin    con   16            'serial transmit pin
Baud    con   84            '9600 baud, 8-bit
S0      con   0             'S0 pin
S1      con   1             'S1 pin

'--------------Main routine-------------------------------

Initialize:
pause 1000
serout sPin,Baud,[cr]               'Prepares StampDaq
                                     buffer

serout sPin,Baud,[cr,"label,light",cr]   'label Excel column
                                          with light

serout sPin,Baud,["cleardata",cr]   'Clear all data
dirs = %00000011                    'make pins 0 and 1
                                     outputs

High S0                             'Set S0 and S1 to 5V to
                                     divide frequency by 100

High S1

again:
count 2, 10, light                  'Count on Pin 2 for 10
ms
serout sPin,Baud,["DATA,light,", DEC light,cr]
                                    'Send data to
                                     StampDaq

goto again
```

I cut an X at the top of the pipette bulb chamber. To maximize system sensitivity, I placed the firefly chamber directly on top of the TSL230 chip.

I used a styrofoam cup to remove residual background light sources. The hardest part of this project was collecting the fireflies. I enlisted the help of my 4-year-old daughter, Jordan. We learned that it's easiest to collect fireflies right at dusk, when it's still light enough to resolve the flying males from the background during periods of no flashing.

I used the BASIC command COUNT to count the number of pulses from the TSL230 (see code above). I set the COUNT command to measure for 10ms, but I determined that the actual resolution was approximately 31ms. The delay comes from the time it takes to execute the rest of the program loop and transmit the data into Excel.

Christopher Holt is a research scientist at NexTech Materials in Columbus, Ohio.

Photography by Dave Mathews

LEGAL AND FREE DIGITAL SATELLITE TV

Use a long-obsolete Primestar dish to pull in a wide world of programming. By Dave Mathews

In 1999, the home satellite TV provider DirecTV acquired one of its competitors, Primestar. The DirecTV and Primestar systems relied on different microwave receiver hardware, so when DirecTV upgraded the 2.3 million former Primestar subscribers to its own mini-dish antennas, it made obsolete all of those Primestar dishes, leaving them out there, unused. This article explains how you can modify one of these old antennas to receive dozens of channels of free television. You won't be able to pull in "premium" channels like HBO, which are encrypted, but for about $50 in additional equipment, you'll be able to receive regional television stations from throughout the U.S. as well as international programming from around the world.

Much of the information we'll need in this project comes from the Free TV webpage at lyngsat.com/freetv/United-States.html. To see all of the channels available, browse this page for the DVB channels. (DVB refers to the Digital Video Broadcasting standard, based on the MPEG2 format.) For the channels that interest you, click the satellite name and then find the channel on the satellite page that comes up. If it's listed on a white background

Pick up a used Primestar dish for under $50 (check eBay) and use it to watch free satellite television.

PARTS LIST:

1. Primestar or DirecPC/DIRECWAY antenna
If you can't find one of these relics in a neighbor's yard, look for them at a marina, where you might see dozens. A DIRECWAY or DirecPC antenna will also work (DIRECWAY, formerly DirecPC, is a satellite broadband service). Attached to the antenna will be one or two filtering modules, Ku-band Low Noise Block (LNB) filters with horizontal/vertical polarity.

2. Free-to-Air MPEG2 receiver
Searching eBay for "FTA MPEG2" will reveal hundreds of new and used units starting under $40. A basic unit is the VIStar 2000, which will support a couple hundred channels carried over ten different bouquets. The Cadillac of receivers is the Pansat 2700a, which supports 5,000 channels. Assuming you buy a cheap receiver, your total hardware cost for this project should be under $50, even using high-quality coaxial cable.

3. 2x4 or 3x4 satellite multiswitch (usually necessary)
You'll need one of these if you're using the older and more prevalent model of Primestar dish, which has two LNB outputs, for horizontal and vertical. If you're using a DIRECWAY/DirecPC or a Primestar dish with just one output, you're good to go as-is.

4. RG6 coaxial cables
You'll need one long cable to go from the antenna to the receiver, plus two short ones if you're using a multiswitch. Get the high-quality stuff.

5. Stable mounting platform for the dish

6. Compass

cables, connect the vertical output from the dish to the 14V port on the multiswitch, and the horizontal output to the 18V port. Then connect the multiswitch's output port to your receiver. This hack lets you switch your receiver between horizontal- and vertical-polarity channels by telling it to receive either 14V or 18V input.

For one LNB output: Rotate the LNB filter so that its sticker aims to the left when you face the focal point of the dish.

2. Choose the satellite you wish to tune in and note its orbit, which is listed on its lyngsat.com page. Intelsat Americas 5 is a good one to start with because it has a very strong signal and it orbits at 97.0° W, positioned centrally to the U.S. (but geostationary over the equator, of course).

Determine the local azimuth and elevation of your target satellite by entering its orbit, along with your own latitude/longitude, into the satellite finder website at satsig.net/ssazelm.htm. Mount the dish to a stable and unobstructed surface outside, and then use your compass to pre-aim the dish towards the satellite. If the bird you want sits below your horizon, you need to pick another one.

3. Temporarily hook up a TV to the receiver using a short coaxial cable, and position the screen so you can see it while you aim the antenna.

4. Set the receiver's Local Oscillator (LO) frequency to 10750, for Ku band.

5. Refer back to your satellite's lyngsat.com page to find a channel bouquet that carries at a frequency of 11,700MHz or higher. On Intelsat Americas 5, a good one to try is at 12,177MHz, run by Pittsburgh International Telecommunications. This bouquet includes a nice variety of channels.

5a. In the bouquet's satellite page listing, note whether it transmits with horizontal or vertical polarity by looking for the "H" or "V" in the first column. Then note the signal rate and forward error correction values, listed in the column labeled "SR-FEC/SID-VPID." Our example, 12177 on satellite IA5, transmits with vertical polarity, and the SR and FEC are 23,000Mbps and 3/4, respectively.

5b. On your receiver, program the frequency of your chosen bouquet, and select its polarity.

5c. Set the bit rate value for the bouquet.

5d. If your receiver requires it, set the forward error correction value for the bouquet.

and its frequency, shown in the first column, is above 11700 (11,700MHz), you can receive this channel — provided your view to the satellite is unobstructed.

Aiming a dish takes a while, so you need to choose one satellite from which to receive your signals (or install multiple dishes, like I have). But each "bird" transmits hundreds of channels of video and audio, some encrypted, some clear. Each satellite broadcasts at dozens of frequencies, and each frequency can pack multiple digital channels simultaneously, which is why the frequencies are referred to as "bouquets." So you can get a lot from one satellite. These days, plenty of good English-language programming comes off the satellites Intelsat Americas 5 and Galaxy 10R, so these are good ones to explore.

Procedure:

1. "Correct" the polarity of the dish.
For two LNB outputs: Using two short coax

When you buy a used satellite dish, it should have a filtering module attached to it, typically a Ku-band Low Noise Block (LNB) filter with horizontal/vertical polarity. If you have an older dish, you'll also need a 2x4 or 3x4 satellite multiswitch, but if you are using a DIRECWAY/DirecPC or a Primestar dish (with just one output) you won't need the multiswitch.

6. Save these values and exit. Open the receiver's signal strength and quality screen, and choose the frequency that you just programmed in.

7. Watching the TV screen and your receiver's display, fine-tune the east/west aiming of the dish to get the maximum signal and quality level. Since this is a digital signal, you need to make small adjustments and then wait a moment after each change. This is a tedious process; it's much more difficult than with analog receivers, which react quickly to your movements. An inline analog signal strength meter can help, but an inexpensive one won't be able to identify the satellite name, and you might wind up zeroing in on a bird that's two degrees away from the one you wanted. If your elevation is off, you can adjust this as well, but note that moving in two axes makes aiming much more difficult.

8. Exit the signal strength screen and initiate a channel search or scan. If the channels you were aiming for show up, then you've aimed the dish correctly.

9. Once you know that you have a good signal from the right bird, program in all of the bouquets that are available on the satellite. Each of these frequencies is carried by a different transponder, which is why they're listed in a column labeled "Freq. Tp" (which stands for "frequency" and "transponder").

10. Initiate a full scan to find all of the channels on the satellite.

That's all there is to enjoying free digital MPEG2 satellite TV. If you get into it, you can buy a smaller, motorized dish to pick up other satellites. Sometimes you'll find surprise "wild feeds" that networks use to distribute prime-time programming to their affiliates for later broadcast. You'll get to see the shows before your friends! A few MPEG2 receivers even support hard drives for TiVo-like operation. The audio feeds you can find are also interesting, but don't expect a professional program guide. This is the Wild West of home entertainment, and it's every programmer for themselves!

Dave Mathews has been tinkering with TVROs since they were first affordable and has a very small array (VSA) of dishes in his backyard. More stories and video clips can be found on his website at davemathews.com.

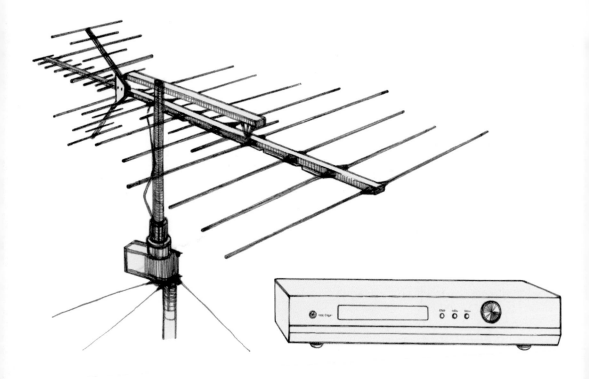

DIGITAL TV ON THE CHEAP

You can get many of the benefits of Digital TV (DTV) for as little as 20 bucks if you're willing to do a little tinkering. By Bob Scott

In most cities, some or all TV stations now broadcast free over-the-air (OTA) DTV signals (see checkhd.com). This signal, and some surplus gear, is the ticket to cheap DTV.

Despite what you might have heard, you do not need an HDTV set to benefit from DTV signals. An external DTV tuner will allow your faithful Philco to receive even high definition (HD) programming, just not at the resolution you see on the pricey plasmas down at the mall.

So why bother? First, even at the lowest DTV resolution (called SDTV, equivalent to current analog TV), the picture will be completely free of snow, ghosts, and other artifacts. If you can receive a digital signal it will be, by definition, perfect. Error-correction protocols in the transmis-

sion assure that the picture you see will be an exact replica of what left the studio.

Second, broadcasters can transmit additional channels called "multicasts" that are only available to digital receivers. These channels typically include news loops, weather radar, and children's programming, in addition to the normal network programming.

Finally, you'll get high-fidelity digital audio in either regular stereo or 5.1 Dolby Digital surround.

I got a refurbished Samsung SIR-T451 DTV tuner to see how hard it would be to make all this work (a used model sells for about $225; new ones are

You don't need a high definition television to get at least some of the benefits of DTV.

about $50 more). The VCR-sized unit has multiple outputs, ranging from radio frequency (RF) all the way up to digital video interface (DVI), making it compatible with a wide range of TVs.

A real hacker, though, goes for a surplus HDTV satellite receiver that incorporates an OTA DTV tuner. These can be had for a couple of bucks at a garage sale, but do your homework. Know exactly what you're looking for, as only certain hardware, firmware, and receiver status configurations will pick up OTA signals when not subscribed to a pay service.

The newsgroup alt.video.digital-tv is a good place to do your research. I've seen old Voom satellite boxes going for $20 on eBay. Although the company went under, the boxes contain a perfectly usable over-the-air HDTV tuner.

(If you're lazy, you can just sit back and wait until March 1, 2007. By that date, all TVs sold in the United States must have a built-in OTA tuner, as decreed by the Federal Communications Commission.)

DTV is generally broadcast on UHF, so you'll also need a good UHF antenna. A set-top loop might work if you're lucky, or you might need a roof-girdling Yagi on a tower if you're out in the sticks.

DTV signals are particularly sensitive to multipath (ghosting), so reception in urban areas can be problematic. If you can get a ghost-free UHF signal on analog TV with your antenna, you'll probably be in good shape for digital, but the only way to be sure is to try.

In my high-rise apartment, I could get only one of the six digital signals in my area, but it was gorgeous. I tried everything from a hunk of wire to an ultra-sophisticated amplified directional antenna, with the same results.

The next experiment was loaning the tuner to a friend in the suburbs of Houston. He connected it to his vintage outside antenna and instantly got every digital channel in the area — much to my annoyance.

I let him keep the tuner while I looked for a better antenna ... or a new apartment.

A little fiddling in my apartment resulted in an interesting discovery: my cable company was pumping HD down the cable, for free. With the proper tuner, I could get six HD channels for just the price of a basic cable hookup, no box or

premium service required. Apparently this is getting more common, but the cable companies aren't exactly taking out any full-page ads about it, preferring to rent you an HD cable box.

Bob Scott is a statistical construct of various consumer electronics marketing departments.

Satellite TV History

Once upon a time, satellite TV was for regional cable companies, not homes. Geostationary satellites 22,300 miles high beamed unencrypted, free-to-air (FTA) broadcasts to Earth. Cable companies tuned them and rebroadcast the content to their subscribers. Home satellite TV debuted in the 1980s with 12-foot dishes that eavesdropped in on these signals. Private backyards started sprouting "BUDs" (Big Ugly Dishes) that moved from horizon to horizon in order to pick up the 24 C-Band analog channels that were spread across 30 different satellites. For a while, anyone who could afford tens of thousands of dollars for receiving equipment would be rewarded with free, unlimited programming, which included premium cable channels and commercial-free feeds from national television networks.

Before long, this equipment dipped below $3,000, and cable networks began losing significant revenue from satellite owners who had stopped subscribing because they were going directly to the source. Something had to be done. So in 1985, HBO and Showtime began scrambling their signals with an encryption scheme called Videocipher II. Under this regime, satellite users had to add a descrambler module in order to watch these premium channels. Of course, people soon figured out how to build their own decoders, or bought black-market versions. It was estimated then that only 10% of satellite dish owners were actually paying for their signals, and nearly all satellite dealers were also selling the illegal decoders.

The mid-1990s saw a closed and fully encrypted system that received digitally compressed signals through a fixed 18-inch microwave mini-dish. Today, DirecTV and EchoStar's Dish Network use this technology. It's still not uncrackable: in December, two men from Tennessee were charged with selling a $250,000 device that made illegal DirecTV decrypting cards.   —Dave Mathews

FULL MAST RECEPTION

Build your own satellite dish mast in three easy steps. By Joe Grand

If you live in an apartment complex and you've read the fine print of your lease, you might know that you're not allowed to attach anything to the railings, walls, or other physical property of the apartment. So, if you plan on setting up satellite TV (for example, DirecTV, Dish Network, or other services), you'll need to use a floor-mounted mast or base to hold your dish in place. The mast must be solid enough to stay put in rain and strong winds and if you bump into it. That way, the dish remains fixed on the network's geosynchronous satellites, orbiting at 22,300 miles above the Earth. The satellites remain in the same location in the sky, so little or no maintenance is required once the dish is aligned.

Some people use off-the-shelf tripod stands for this, but they're hard to secure unless you bolt them down or place a bunch of sand bags around them. This is not very aesthetically pleasing out on a patio, so let's build our own!

Step 1: Obtain the Parts

Visit your local home or hardware store and obtain these two items:

1. A heavy concrete umbrella base like the ones you commonly see at sidewalk cafes. I used the "Umbrella Base White 7-inch" from Home Depot, UPC #96968-60699 (which, despite the name, is 14" in diameter).

2. A 6' length of 1½" diameter fence piping. For me, the total cost of the materials, including tax, was just $22 — well worth the money!

Step 2: Place the Base

Drag, lift, or otherwise move the umbrella base onto your apartment's patio, deck, or wherever else you desire. Be sure to wear shoes during this step, as the base can easily crush toes. Place the base in a spot where your antenna will have an unobstructed view of the southern sky, where geosynchronous satellites ring the equator.

Step 3: Host the Post

Conveniently, the 1½" diameter of the fence piping fits snugly into the umbrella base and also matches the mounting clamp of the satellite dish. I'm using a Phase III 18x20" Triple LNB dish manufactured for DirecTV, but most other dishes have similar mounting hardware.

Simply insert the fence piping into the umbrella base and tighten the base's setscrew to secure the pipe. Tighten the screw hard with a wrench, so it doesn't wobble or become loosened or misaligned over time.

With the post in place, verify that the post is vertically true, so you can accurately aim the dish. Use a standard level and just place it vertically along the post; tweak the post and concrete base as necessary. Two possible ways to even up the post alignment are to place a shim or two underneath the base, or to rotate it left or right by a few degrees at a time. When your post is firmly attached and vertically true, you can attach your dish and align it in accordance with your satellite receiver directions.

Step 4 (Optional): Watch TV

You're done! Now that your mast is complete, your dish is aligned, and all of the cables are routed properly to your TV or DVR, it's time to sit back and melt your brain with the multitude of fantastic television stations available on the satellite networks. If you're anything like me, you'll surely take much advantage of your newly built creation.

Joe Grand (joe@grandideastudio.com) is the president of Grand Idea Studio, Inc., a product research, development, and licensing firm. He is the author of the books *Game Console Hacking* and *Hardware Hacking: Have Fun While Voiding Your Warranty*.

PM—▶

Alarm ▶

FM   88 92 96   104   108   MHz
AM   54 60 70 80 100   140 170   KHz

AM - FM Electronic Clock Radio/Cassette Player **SOUNDESIGN**

A crop of makers'
materials is ripe for the
picking inside this old
radio.

DUMPSTER CORNUCOPIA
Reusing components from discarded electronics. By Thomas Arey

Photography by Thomas Arey

Surface mount components have revolutionized all aspects of the electronics industry. However, they have been somewhat of a mixed blessing for electronics hobbyists. Homebuilders, especially those involved in the radio hobby, are experiencing a shortage of some through-hole (traditional leaded) components.

This parts shortage problem has happened before. The move from vacuum-tube-based technology to transistors, and the subsequent advance of integrated circuits over discrete semi-conductors, both sent electronics hobbyists in search of rare parts to build fun and exciting things.

Closeout and surplus dealers are always a possibility, as are online auctions. But I enjoy using the time-honored technique of salvaging parts

from discarded electronic devices. Most households relegate broken or obsolete electronics to the trash because the industry has advanced to the point that new devices can often be purchased for less than the cost of repairing the original unit. The politics of this throwaway economy are beyond the scope of this article, but it remains both a preference and a challenge, for me personally, to turn the parts of discarded electronics into new projects.

I do this not only with electronics discarded in my own household, but also with devices scrounged from friends and neighbors. I even resort to a bit of casual dumpster diving when I am out walking my dogs on local trash days. It is not uncommon for my walks to turn up discarded radios, televisions, VCRs, children's toys, and

Fig. 1

Fig. 2

Fig. 3

This simple radio took less than an hour to bring down to a nice quantity of useful parts for future projects. More complex devices, such as a VCR, can yield ten times the number of components, not to mention dozens of useful hardware pieces such as self-tapping screws.

even antique computer systems. All such devices can be treasure troves of useful and even rare parts.

Here are a couple of important warnings, though, before you join me in the quest for reusable electronics parts. First, be aware of how discarded electronics can cause you bodily harm — everything from sharp edges to imploding TV picture tubes. Handle all found items with caution and care. Also, be aware that some capacitors can hold a dangerous charge even in an unenergized, discarded device. Again, be cautious and treat every device as a live circuit, even after you have cut the line cord off and added it to the box under your workbench.

Finally, it's very unwise to pick through another person's trash without asking permission. Some people see this as a violation of their privacy.

As a test case for parts scrounging, let's take a look at this Soundesign model 3833 AM-FM clock radio/cassette player that I recently found in a trash can a few blocks from my home.

Initially, opening the case and dismantling the radio down to the circuit board level provides me with (Fig. 1): a line cord, 110-to-12-volt transformer, 3-inch speaker, a small motor from the cassette deck, 9-volt battery connector, recording head, and assorted knobs, belts, and hardware.

An examination of the circuit board shows many useful components (Fig. 2). You can strip a board such as this to whatever level you choose. I usually leave the standard capacitors and resistors but remove the semiconductors, ICs (if not proprietary), and any variable capacitors and potentiometers.

Final desoldering adds to my parts drawers (Fig. 3): an AM ferrite antenna, LED display, trimmer capacitor, potentiometer, variable capacitor, two multi-position switches, six slug-tuned inductors, voltage regulator IC, op-amp IC, two NPN transistors, and six diodes.

While not a perfect solution to the growing scarcity of some electronic parts, giving discarded electronics new life in projects can be a lot of fun.

T.J. "Skip" Arey N2EI has been a freelance writer to the radio/electronics hobby world for over 25 years and is the author of *Radio Monitoring: A How To Guide*, Paladin Press.

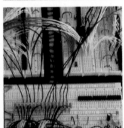

THE ROCKETMAN'S GARAGE

Ky Michaelson, a.k.a. The Rocketman, hails from an illustrious line of makers. On his family tree can be found inventors of the motorcycle transmission and clutch, the rotary lawn-mower blade, the flip-top aspirin box, and the oxygen mask as used in commercial aviation.

As a child dealing with dyslexia, Michaelson struggled in school, but discovered a natural affinity for building things. One day in math class his teacher discovered that he'd hollowed out his textbook and built a crystal set inside of it (the earpiece wire was concealed through his shirt sleeve). She was awestruck by his ingenuity (in addition to being pissed at him for destroying his book), and it was this ability to dazzle with his creations that sent him on a life-long quest to build more impressive machines.

Michaelson is forever hungry for power and speed. In his Gyro Gearloose world, he's never met a moving vehicle he didn't think would go better with a big-ass rocket engine bolted to its backside. And, at this point, there aren't too many things he hasn't "rocketized," from motorcycles and go-karts to tricycles and kiddie scooters. And then there are the rocket-powered toilets and barstools. See more of Rocketman's crazed creations at the-rocketman.com.

—*Gareth Branwyn*

Photograph by Chad Holder

1. Replica of an early Michaelson motorcycle, originally built by Michaelson's great uncles of Michaelson Motor Company. 2. Rocket-powered go-kart. His kart holds the world record of 252 mph. 3. The payload section of Michaelson's Civilian Space Exploration Team's GoFast rocket, which, on May 17, 2004, became the first amateur rocket in space. 4. Dubbed the "SS Flusher," this rocket-powered toilet came about when somebody overheard Michaelson saying that a commode was one of the few things he hadn't made into a rocket. The bowl donor worked for the prison system and donated it on the q.t. 5. This hybrid (N_2O/plastic) trike was built for Michaelson's son, Buddy (age 6). Not surprisingly, "Rocketman" is his middle name. No, really. Legally. 6. An H_2O_2 rocketpack. Michaelson's trying to convince his wife Jodi to take it for a spin, which would make her the first woman to fly a jetpack. 7. An original 1912 Michaelson motor that the Rocketman found at a flea market in Davenport, Iowa. He'd been in search of one for 30 years. 8. A rocket-powered bicycle Michaelson built for Jodi using a Mongoose frame. 9. A twin-engine drag bike, built on a BSA frame. 10. A steering "engine bell" off an Apollo rocket. Michaelson has over 275 space artifacts in his home in the Minneapolis suburbs.

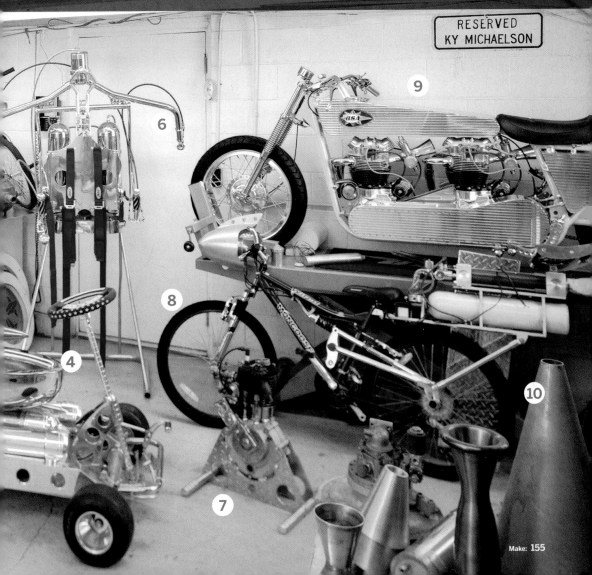

RESERVED
KY MICHAELSON

Welcome to the first installment of AHA!, a page of MAKE's favorite puzzles, selected by Michael Pryor. When you're ready to check your answers, visit makezine.com/05/aha.

The Rope Bridge (Easy)

Four people need to cross a rickety rope bridge to get back to their camp at night. Unfortunately, their only flashlight has just 17 minutes of juice left. The bridge is too dangerous to cross without a flashlight, and it's only strong enough to support two people at any given time.

Each of the campers walks at a different speed. One can cross the bridge in 1 minute, another in 2 minutes, the third in 5 minutes, and the slowpoke takes 10 minutes to cross. How do the campers make it across in 17 minutes?

Yarrr Maties (Sinister)

Five pirates discover a chest full of 100 gold coins. The pirates are ranked by their years of service, Pirate 5 having five years of service, Pirate 4 four years, and so on down to Pirate 1 with only one year of deck scrubbing under his belt. To divide up the loot, they agree on the following:

The most senior pirate will propose a distribution of the booty. All pirates will then vote, **including the most senior pirate**, and if at least 50% of the pirates on board accept the proposal, the gold is divided as proposed. If not, the most senior pirate is forced to walk the plank and sink to Davy Jones' locker. Then the process starts over with the next most senior pirate until a plan is approved.

These pirates are not your ordinary swashbucklers. Besides their democratic leanings, they are also perfectly rational and know exactly how the others will vote in every situation. Emotions play no part in their decisions. Their preference is first to remain alive, next to get as much gold as possible, and finally, if given a choice between otherwise equal outcomes, to have fewer pirates on the boat.

The most senior pirate thinks for a moment and then proposes a plan that maximizes his gold, and which he knows the others will accept. How does he divide up the coins? And what plan would the most senior pirate propose if the boat had 15 pirates instead of just five?

Michael Pryor is the co-founder and president of Fog Creek Software. He runs a technical interview site at techinterview.org.

IR Remote Control Protocol

Get an infrared remote to turn your room lights on and off.

By Andrew "Bunnie" Huang

THEORY

Back when I was an undergraduate at MIT eight years ago, my fraternity had a loft in every room. This suited me fine, but there was no way to turn the light switch off from up there when it was time for some sleep, and I certainly did not want to trundle across a floor land-mined with DIP ICs, papers, and tools, then climb up the precarious ladder to my roost without the benefit of sight.

Meanwhile, I also had a Sony stereo system with a remote control that sported a TV power button, but I had no TV. And so, like peanut butter and chocolate, two worlds came together. I decided it was time to put that derelict TV button to good use: switching the lights off at night. Doing this meant making a light switch that could understand signals from an infrared remote control — specifically, one that used Sony's IR protocol, since different manufacturers do infrared in different ways.

How IR Remotes Work

At a physical level, an IR link consists of an infrared LED on the remote, which typically emits light at a wavelength of around 940-980nm, and an infrared photodiode detector on the main piece of equipment. The detector is covered by an IR-selective filter, usually a deep red piece of plastic, that helps block stray ambient light. The detector turns incident photons into electrons that can be easily sampled and amplified.

A naive implementation of an IR link would be to transmit a beam of infrared light to represent a "1" or On condition, and use the absence of signal to indicate "0" or Off. While this can yield a functional link under controlled conditions, there are several reasons why it won't work at home. Many household light sources emit energy at wavelengths that excite the IR detectors typically used in home electronics. This would make it difficult or impossible for the receiving device to distinguish intentional IR signals from stray noise in the red/infrared range — for example, sunlight reflecting off a Coke can. This problem is known as spurious pulse rejection.

> "I decided it was time to put that derelict TV button to good use."

Another problem is that remote controls need to work in a wide range of ambient light conditions, between bright sunlight and pitch dark. Without some encoding scheme for the signal, it's difficult to build a receiver that can filter out such a large possible range of ambient light and still be sensitive to the data it needs to receive.

The solution that most IR remotes use is to modulate the IR light with a carrier frequency. My Sony remote, for example, used a carrier frequency of 40kHz, which is fast enough and regular enough that it is unlikely to be confused with random fluctuations and reflections. This frequency also allows noise from ambient light hitting the IR photodiode to be largely filtered out by a relatively small coupling capacitor.

Many IR receivers add more complex circuitry to comb out spurious signals prior to bit detection. These include tuned filters that select (bandpass) and reject (notch) particular frequencies, and synchronous detection circuits, which multiply the signal by the carrier frequency you are looking for. Because sine waves at unrelated frequencies cancel each other out overall, this multiplication results in only the carrier frequency and its harmonics being left with a nonzero result.

Detector modules built to support the 40kHz standard also make great sensors for object detection in robotics, because of their robust handling of spurious signals.

IR Protocol

Using a carrier frequency will solve the general problem of spurious signals, but you still need a way to encode data for transmission. Sony remote controls use what's called pulse-duration modulation, where the value of a bit is communicated by the duration of a pulse of the carrier frequency — not

unlike the dashes and dots of Morse code. In Sony's scheme, 1 is represented by a 1,100-microsecond (μs) burst of 40 kHz light, and 0 is represented by a 550μs burst.

Data streams over the IR link in LSB-first order (LSB = least significant bits), with an extended start bit of 2.4-millisecond (2,400μs) duration to indicate the beginning of a signal transmission. The extended start bit allows IR receivers to save power by running idle during periods of no transmissions, then have enough time to wake up and watch when there's a start signal.

Each key-press event on a remote sends an IR data packet consisting of two parts: a 7-bit command followed by a 5-bit device address. The command encodes the identity of the key that was pressed, and the device address specifies what type of appliance the key press was meant for. For example, the power button is key number 21, a TV is at device address 1, and a CD player is at device address 17.

False IR Signals

A. Bright, shiny can logo

B. Path of reflected light

C. On/off points for
false binary, from
reflections off can

D. False binary signal

E. Equivalent bar code

IR protocol is designed to prevent the possibility of spurious signals, but in the noisy, light-filled world, there are unlikely instances where an IR receiver might detect something that looks like binary data.

Consider this pathological case (of soda): Someone in the room moves a shiny, printed aluminum can near a window in the sun. Sunlight reflects off the can, and bounces directly into the IR receiver. The path of incidence of the beam hitting the sensor also happens to travel across a two-color logo. One of the logo's colors absorbs IR, while the other color reflects it. As a result, the sensor gets a short blast of fast-pulsing IR.

With a sensor module that combs everything it sees for binary patterns, this random reflection might appear to be a signal from a remote. Hopefully it doesn't look like the dreaded "self-destruct" button.

It's also possible to intentionally create reflective patterns that you shine light across and read as data. This is how bar codes work.

PRACTICE

Knowing this background, it's not difficult to build an IR receiver that operates a relay for switching AC devices like lamps on and off. It will be a small box with an IR detector peeking out the side, attached to the middle of an ordinary extension cord.

The Brains

The brains of our IR receiver light switch will be an 8-bit microcontroller chip (a PIC16F84A), running my adaptation of code originally written by R. Dunbar Poor that enables it to perform remote control tasks. This code is republished with his permission at web.media.mit.edu/~ayb/irx/irx2. In addition, our circuit also needs a crystal oscillator and a voltage regulator, to support the PIC.

The Eyes

For our system's "eye," we can use a standard IR demodulator such as Sharp Electronics' GP1UD28YK, available from Digi-Key for about $1. These wonderful modules perform the tasks of IR signal detection, amplification, limiting, filtering, demodulation, integration, and digitization, all in a very convenient 3-pin package. The Sharp IR module connects easily to a PIC; you just wire up its power and ground pins, then connect its output pin to one of the PIC chip's inputs. I chose to connect to the chip's port B4 (pin 10).

The Switch

To actually switch the lamp on and off, we can use any 5V-activated power relay with sufficient current rating for the lamp; one example is the Omron G6B-1114P-US-DC5. Unfortunately, the PIC's output current of 25mA isn't quite enough to activate most relay coils, which need something in the high tens or low hundreds of milliamps (the Omron relay needs 40mA). So we use a transistor to boost the current.

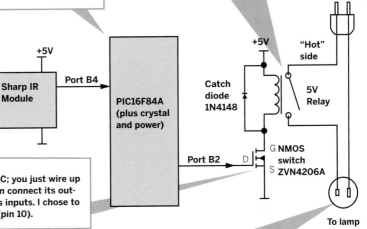

The Booster

To give the PIC output a boost, we can use a transistor, such as the Zetex ZVN4206A; it has a continuous drain current rating of 600mA and a low gate-source threshold voltage (<3V), which means the PIC output will be able to drive it. We can wire up the entire circuit on a PICProto18 prototyping board; the diagram at right shows how the main components connect.

VERY IMPORTANT: Once your switch is built, before you plug in for the first time, verify all your connections with an ohmmeter; there should be near-infinite resistance between all high-voltage lines and every other node in the circuit. If so, it's safe to plug the cord in, plug your lamp (or anything else) into the cord, and click-on, click-off. Pleasant dreams!

The Plug

After you've built the remote control switch box, you'll need to hook it into the extension cord that runs between the wall socket and the lamp. Take an extension cord, cut its "hot side" conductor (the narrower, right side of the plug), and splice the switch's relay terminals in-line into this conductor. Be careful when wiring up the 120VAC lines. As you well know, dangerous voltages will flow through these wires, so you need to use grommets to prevent the edges of your case from cutting into the wire. It's also a good idea to tie strain-relief knots in the cords, to prevent sudden tugs from pulling the wires out of your board. For good measure, seal any exposed high-voltage wires with a dollop of hot glue.

Andrew "Bunnie" Huang lives in San Diego. He likes to hack.

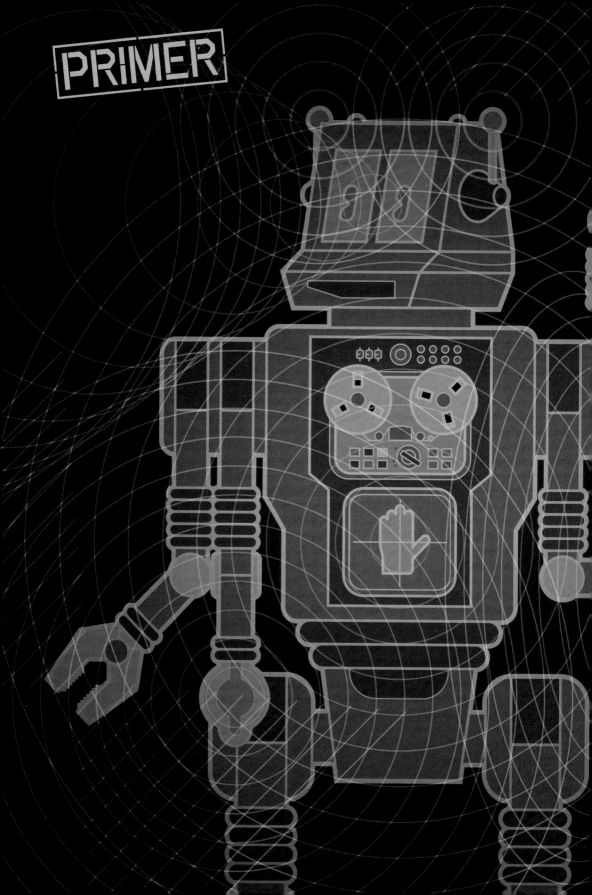

PRIMER

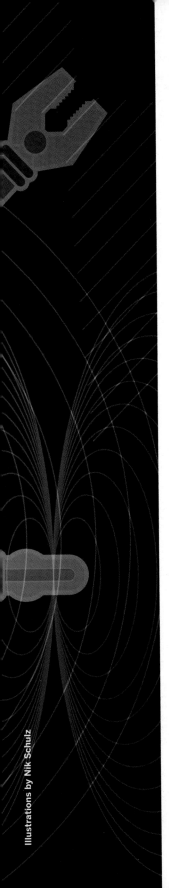

Illustrations by Nik Schulz

Sensor Interfaces

How circuits communicate with the outside world. By Tom Igoe

A typical home appliance contains several sensors, from the switches and dials you interact with to the temperature sensors and limit switches that keep your coffee pot from overflowing or setting the house on fire when the last cup is gone and the heater's left on. Sensors are central to a hardware hacker's vocabulary. The more of them you know about, the more responsive your devices can be.

What follows is an introduction to the characteristics of most sensors, so you know the range of possibilities. Later, I'll show how to use a few specific sensors.

This article assumes you've got a basic understanding of how electricity works. If you don't, check out Forrest M. Mims' book, *Getting Started in Electronics*.

Types of Sensors

Electronic sensors convert other forms of physical energy into measurable electrical properties.

All sensors can be grouped in two major categories: **digital sensors**, which output one of two possible states, on or off, and **analog sensors**, whose outputs change continuously with the changing physical energy that they read.

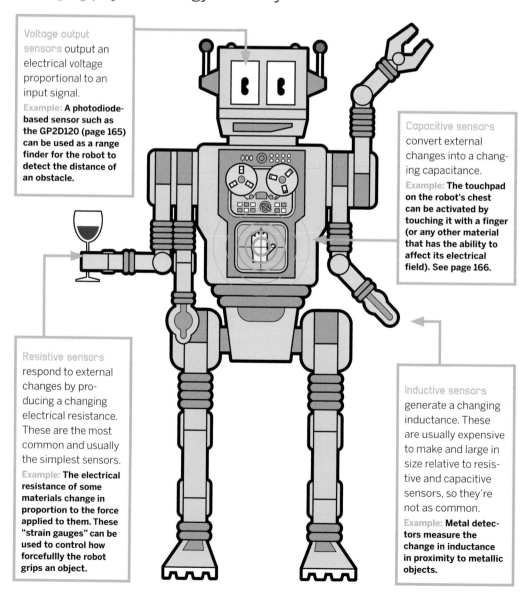

Voltage output sensors output an electrical voltage proportional to an input signal.

Example: **A photodiode-based sensor such as the GP2D120 (page 165) can be used as a range finder for the robot to detect the distance of an obstacle.**

Capacitive sensors convert external changes into a changing capacitance.

Example: **The touchpad on the robot's chest can be activated by touching it with a finger (or any other material that has the ability to affect its electrical field). See page 166.**

Resistive sensors respond to external changes by producing a changing electrical resistance. These are the most common and usually the simplest sensors.

Example: **The electrical resistance of some materials change in proportion to the force applied to them. These "strain gauges" can be used to control how forcefullly the robot grips an object.**

Inductive sensors generate a changing inductance. These are usually expensive to make and large in size relative to resistive and capacitive sensors, so they're not as common.

Example: **Metal detectors measure the change in inductance in proximity to metallic objects.**

Sensor Interfaces

In order to use sensors, you have to know how they interface with your computer, microcontroller, or circuit. Sensor devices typically combine a sensing component with filter circuits and amplifiers to convert the sensor's electrical changes into an electrical signal that's usable by microcontrollers and other electronic devices.

There are a handful of common types of sensor interfaces. Some can be used directly in an output circuit, and others need to be connected to a computer or microcontroller.

The simplest sensor devices output a changing DC voltage. An **analog sensor** will output a voltage that changes continuously across its output range, and a **digital sensor** will output either the top of its range or the bottom. For example, the Sharp GP2D120 analog sensor outputs a changing voltage, from 0 to about 3 volts, that corresponds to a person or object's proximity to the sensor. The Quantum QT113H sensor, a digital sensor, outputs 0 volts when a person is in contact with the sensor, and 5 volts when no one's in contact with it.

Pulse width output sensor devices will generate a series of digital pulses that change with the energy they're sensing. For example, a Taos TSL230 sensor converts the intensity of light received into a changing frequency of pulses. You'll often see the term PWM, or pulse width modulation, to refer to this form of output.

More complex sensor devices output digital data serially, either using an asynchronous serial protocol that's common to most personal computers, or using a synchronous serial protocol.

Asynchronous serial devices work just like any other device that connects to a computer's

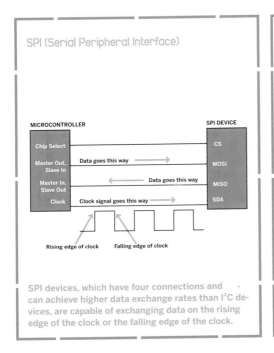

SPI (Serial Peripheral Interface)

SPI devices, which have four connections and can achieve higher data exchange rates than I²C devices, are capable of exchanging data on the rising edge of the clock or the falling edge of the clock.

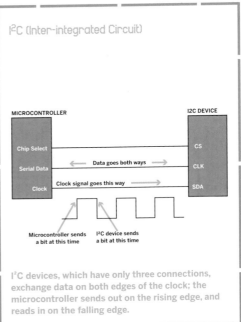

I²C (Inter-integrated Circuit)

I²C devices, which have only three connections, exchange data on both edges of the clock; the microcontroller sends out on the rising edge, and reads in on the falling edge.

serial port. They have three connections: one for transmitting data, one for receiving data, and one that's a common ground. Some can be connected directly to a PC's serial port. Others can be connected through a signal converter like Maxim's MAX232 chip. Once you know the sensor device's data protocol, you can read the data right from your PC.

Synchronous serial devices don't have their own oscillators, so they need an oscillator (clock) signal from another device. The usual way to use a synchronous serial sensor is to connect it to a microcontroller, which then interfaces to an output device, desktop computer, or network. There are two common flavors of synchronous serial communication: SPI and I2C.

SPI (Serial Peripheral Interface) devices have four connections: clock, data output (sometimes called MISO, or Master-In-Slave-Out), data input or MOSI (Master-Out-Slave-In), and a chip select pin. Because it's possible to connect several SPI devices in parallel on one data bus, the chip select pin is used to tell a chip to respond to or ignore incoming data.

I2C (Inter-integrated Circuit) devices have only three connections: clock, data, and ground. The

microcontroller that controls an I2C device shares the data line with the device. Typically, the controller sends a bit of data when it changes the clock pulse from low to high, and the controlled device sends data when the clock pulse goes from high to low.

Most microcontrollers have I2C and SPI interfaces built in, and software libraries for using these interfaces so you don't have to know a whole lot about SPI or I2C, as long as you know the format of the sensor device's data and which pins to connect.

Every sensor has a data sheet from its manufacturer that explains its physical and electrical characteristics, and its operating conditions. If you're lucky, sample circuits and applications will be included as well. When working with a new sensor, download the data sheet first and have it available to figure out the details.

Examples: Voltage Output Sensors

Sensors that output a changing voltage are generally the simplest sensors to use, both with a microcontroller and with other circuits. Following are two examples: the Sharp GP2D120 infrared distance-ranging sensor (page 165), which detects an object and reports its distance to the sensor as a changing voltage from 0 to 3 volts, and the Quantum QT113H capacitive touch sensor (page 166), which reports the presence of a person's touch by changing its output from 5 to 0 volts. Both sensors are available from a variety of retailers. I get my Sharp sensors from Acroname (acroname.com) because they also sell a pre-made connector cable for these sensors, which can be a pain to make yourself. Digi-Key (digikey.com) sells the QT113H sensors.

If you're interested in using sensors, you should learn how to use microcontrollers as well (*see Primer in MAKE Volume 04 for a tutorial on microcontroller programming*). Microcontrollers are the small, simple computers that drive everything from your car engine to your light switches. The most common way to use a sensor is to connect it to a microcontroller that reads the sensor and controls some form of physical output device based on the sensor's readings. Parallax's (parallax.com) Basic Stamp is an excellent microprocessor development platform for sensor-based application. The company also sells a wide variety of sensors (search for "sensor" on Parallax's website).

Thermocouple: Simple Voltage Output Sensor

Voltmeter — Measuring Junction (Hot) — Reference Junction (Cold) — Heat Source — Dissimilar Metal Wires

Thermocouples are inexpensive and widely used voltage-based temperature sensors. In 1831, T. J. Seebeck discovered that two dissimilar metals will produce a voltage that's proportional to the temperature difference between two junctions of the metals.

GP2D120 Infrared Distance-Ranging Sensor

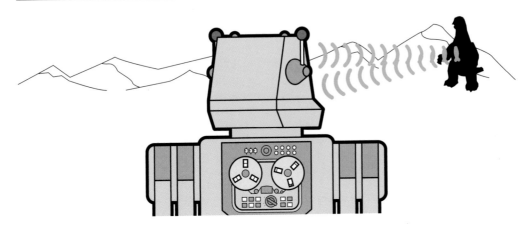

The GP2D120 measures distance by bouncing a beam of infrared light off a target and reading the reflection with an infrared photodiode that converts the incoming light to a variable voltage. The closer the target, the brighter the reflected light, and the higher the output voltage. It has a range of 1.5 to 12 inches.

This sensor comes from a family of IR ranging sensors from Sharp. Some of them have an analog voltage output, like the GP2D120, and others have a discrete output; that is, they output 0 volts until the target crosses a threshold, and then they output 5 volts. The GP2 sensors come in a range of distance sensitivities, from this one, which has the shortest range (4 to 30cm), to the GP2Y0A02YK, which has an analog range of 20 to 150cm.

The connections for the GP2D120 are about as straightforward as you'll ever see in a sensor. There are three pins: a voltage supply pin that you connect to a 4-6V DC source; a ground pin that you connect to ground; and an analog voltage-out pin that you connect to your microcontroller or sensing circuit. Place a 10-microfarad electrolytic capacitor between the voltage supply and ground as close to the sensor as possible to minimize power fluctuation when the sensor draws power.

You can build a circuit to sample and hold the sensor's voltage, or to convert it to a frequency, but I prefer to do that work using microcontrollers. Most microcontrollers have an onboard analog-to-digital converter (ADC) connected to one or more of their pins, to which you can directly connect the GP2D120's output pin. This sensor's output is not linear, but follows a curve, so using a microcontroller is an easy way to make the output linear, with a little math. See the sensor's data sheet (acroname.com/robotics/parts/R146-GP2D120.html) for the details of the output curve.

The Sharp GP2D120, attached to an AVR ATmega8 microcontroller. The schematic shows only the sensor's connections. Connect to your own circuit or microcontroller wherever it can take 0-3V DC input.

Example code to use the GP2D120 on the following microcontrollers can be found at **makezine.com/go/sensor**.
» NetMedia's BX-24 microcontroller
» Microengineering Lab's PicBasic pro for the PIC microcontroller
» Wiring/Arduino syntax for the AVR ATMega8

QT113H Capacitive Touch Sensor

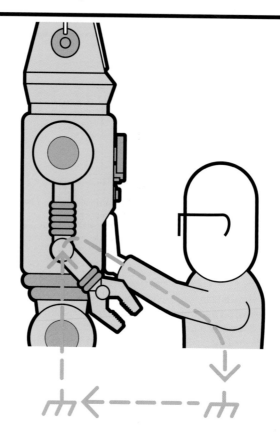

Objects and people always carry an electric charge, and every object or person has a slightly different charge. When two objects or people touch, they exchange electric charge, so that they end up with an equal charge. Capacitive sensors detect charge when we touch an object they're attached to.

Since humans and other animals are basically bags of electrolytes in water, we cause changes in the electrical fields we pass through. The Quantum Technologies QT113H touch sensor measures the change in an electrical field and outputs a voltage when the change crosses a threshold. It responds to human touch, and can read a change through several millimeters of non-conductive material, so it's a great sensor to use when you want to make a fabric switch, or a stuffed animal that responds to touch, or to make just about any surface responsive.

Capacitive sensors take advantage of the fact that objects and people carry an electric charge, and that every object or person has a slightly different charge. When two objects or people touch, they exchange electric charge, so that they end up with an equal charge. Charge-transfer sensors detect charge when we touch an object they're attached to. A conductor is attached to the object, or embedded

in it, and when a person comes close to the sensor, charge is transferred, and the sensor reads it.

To use the sensor, you need to attach a piece of metal to one of its input pins. Copper mesh makes a great sensing electrode for this sensor, and it's relatively easy to solder a wire to. It can also be embedded under the surface of a fabric, wood, or plastic object fairly easily. The wire you use to connect the sensing electrode to the QT113H is also sensitive, unless you shield it. Shielded wire is easy to come by — most computer cables are shielded, and old ones can be chopped up to do the job. The shield is the metal foil or wire braid that surrounds the inner wires. Attach it to the ground of your circuit board. The end nearest the sensing electrode can hang free. Any wire that's outside the shield will sense touch, along with the electrode itself.

The sensor itself comes in two different physical packages, a SOIC chip or a DIP chip. The former is a

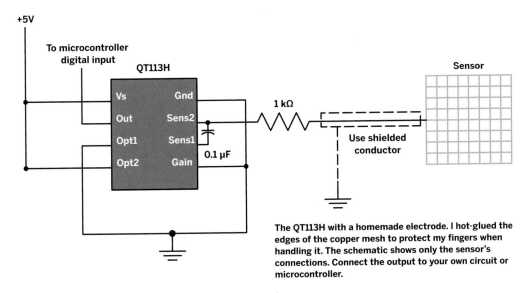

+5V

To microcontroller digital input

QT113H

Vs	Gnd
Out	Sens2
Opt1	Sens1
Opt2	Gain

0.1 µF

1 kΩ

Use shielded conductor

Sensor

The QT113H with a homemade electrode. I hot-glued the edges of the copper mesh to protect my fingers when handling it. The schematic shows only the sensor's connections. Connect the output to your own circuit or microcontroller.

surface-mount chip, very small, difficult to solder to a printed circuit board by hand, and impossible to use directly with a solderless breadboard. I prefer the DIP models for hobby projects. The top of a DIP chip is usually marked with a notch, and Pin 1 is usually marked with a dot. The eight pins of the chip are counted in a U-shape from the top left to the top right.

The pins of the QT113H are as follows:

Pin 1: Supply voltage (V). Connect to 5V DC.
Pin 2: Output. Outputs 0V if there is contact, 5V if not.
Pin 3: Option Pin 1. The option pins control the sensor's reaction to prolonged contact. The sensor normally outputs 5V on contact, and outputs 0V after a set timeout. By changing the states of the option pins, you can set the timeout to 10 seconds, 60 seconds, infinite, or you can set the sensor to toggle every time it's touched. For details, see the sensor's data sheet (qprox.com/downloads/datasheets/qt113_105.pdf). I set the option pins for an infinite timeout by connecting Option Pin 1 to ground, and Option Pin 2 to 5V.
Pin 4: Option Pin 2.
Pin 5: Gain. Connect this to 5V for high sensitivity, or to ground for low sensitivity. In this example, I connected to ground to get more consistent results.
Pin 6: Sensing Pin 1. Connect this to Sensing Pin 2

using a capacitor. Quantum recommends 10 nanofarads, but in this example, I used a 0.1-microfarad capacitor, which worked well.
Pin 7: Sensing Pin 2. Connect this to Sensing Pin 1 using a capacitor, and to the sensing electrode. Put a 1-kilohm resistor in series with the electrode.
Pin 8: Ground. Connect this to the ground of your circuit.

With the QT113H connected like this, it will output 0 volts whenever it's touched, and 5 volts when it's not. You can connect it directly to a relay or transistor to control a motor circuit, or to a microcontroller as part of a more complex system.

If you understand the basic types of sensors and the sensor interfaces shown here, you can figure out how to use most sensors on the market. In following articles, I'll explain how to use a variety of more complex sensors. For a list of sensor-related resources, visit: tigoe.net/pcomp/resources.

> Example code to use the QT11x family on the following microcontrollers can be found at **makezine.com/go/sensor2**.
> » NetMedia's BX-24 microcontroller
> » Microengineering Lab's PicBasic pro for the PIC microcontroller
> » Wiring/Arduino syntax for the AVR ATMega8

Tom Igoe teaches the sensor workshop class and heads the physical computing area at the Interactive Telecommunications Program at New York University.

HowToons
.com

BY SAUL GRIFFITH, NICK DRAGOTTA & JOOST BONSEN '05
THESE LIME GREEN LEAKS INSPIRED BY SCIENCE HULK DANIEL ROSENBURG!

HEE.. HEE..HEE..

NO LICKERS

DON'T YOU READ? IF YOU MUST STAY THEN PLEASE CLOSE THE DOOR.

WHY ALL THE THEATRICS?

I'M **BENDING LIGHT!**

CAREFULLY CUT A 1/2" (1.5CM) HOLE IN THE LOWER THIRD OF THE BOTTLE.

PUNCH OR CUT A HOLE IN A PIECE OF DUCT TAPE. ALIGN THE TWO HOLES AND STICK THE TAPE TO THE BOTTLE.

FIRST, POUR SEVERAL DROPS OF MILK INTO THE BOTTLE. THEN FILL IT UP WITH WATER. BE SURE TO PLUG THE HOLE WITH YOUR FINGER. THE MIX SHOULD BE CLOUDY BUT SEE-THROUGH.

ONCE THE BOTTLE IS FULL, CORK THE TOP (OR SCREW-ON THE CAP) AND REMOVE YOUR FINGER. THE WATER WILL STAY IN THE BOTTLE UNTIL MORE AIR IS LET IN THROUGH THE TOP.

TO GET A STEADY BEAM OF LIGHT YOU CAN USE A FLASH-LIGHT OR LASER POINTER*. IF USING THE FLASHLIGHT, TAPE IT DOWN SECURELY INTO THE BACK OF A SHOE BOX, AND TURN IT ON. POKE A SMALL HOLE WHERE THE LIGHT IS MOST CONCENTRATED.

TWIST THE BOTTLE AND ALIGN THE BEAM SO IT SHINES THROUGH THE BOTTLE AND OUT THE DUCT TAPE HOLE.

From the forthcoming book *HOWTOONS Book 1* by Saul Griffith, Nick Dragotta, and Joost Bonsen. Published by arrangement with ReganBooks, an imprint of HarperCollins Publishers, Inc.

MakeShift

By William Lidwell

This issue's MakeShift challenge is brought to you by Woody Norris. Woody will also participate in the analysis and winner selection.

The Scenario: You and Woody have been asked to evaluate a new bank vault with a time-release mechanism that cannot be overridden. Given the size of the vault and the ¾" vent tube in the ceiling, a person could survive no more than 24 hours before the CO_2 level becomes lethal. Woody claims that with its current contents (left inside by the construction crew), he could survive 48 hours. You feel like he is challenging you so you agree with him, even though you don't see an immediate solution. The next thing you hear is Woody saying, "OK ... see you in 48 hours!" The vault door closes and the lock is set to 48 hours before you can do anything.

The Challenge: Create a makeshift solution to stay alive until the bank vault unlocks. The lights and water cooler have electricity, but that's it. Good luck!

Supply List:

A small table with a desk telephone (line is dead)

1 water cooler with paper cups

4 emergency lights with battery backup

1 steel cable (about 10 feet)

1 garden hose

1 fire extinguisher (CO_2)

Toolbox containing hammer, pliers, screwdriver, pocket knife, electrical tape, chalk-line reel, yardstick, wire, and dirty rags.

Send a detailed description of your MakeShift solution with sketches and/or photos to *makeshift@makezine.com* by April 15, 2006. The most plausible and most creative solutions will each win MAKE T-shirts and a SwissMemory USB Victorinox 512MB. For rules and solutions to previous MakeShift challenges, visit *makezine.com/makeshift*.

William Lidwell is a consultant with Stuff Creators Design Studio and co-author of the book *Universal Principles of Design*.

Photograph by Jill Butler and Christopher Hujanen

The best tools, software, gadgets, books, magazines, and websites.

1. The Perfectly Balanced Kitchen Knife

Wüsthof 10-Piece Classic Block Set

$400 wusthof.com

▪ Before discovering the awesome blade of the Wüsthof knife, I avoided recipes that called for cutting up fresh ingredients. Cutting things is more fun when you're doing it with a high-carbon specialty knife that was crafted with a laser beam.

Now, armed with my Wüsthof Classic knife set, I've been tossing freshly cut veggies into omelets, creating my own fruit trays, and leaving precut loaves of bread for the culinary patzers.

The knives come with a sharpening steel, and if you use it correctly, they should last longer than you will. I sharpen them every three or four uses, and it's working out fine. The entire set fits into a nice wood block for you to store away, and you get a lifetime warranty that covers defects in craftsmanship.

—Matthew Russell

2. Turn Your Records into MP3s

ART Pro USB Phono Plus $99

artproaudio.com

▪ You have "Kung Fu Fighting" on vinyl, but how do you get that into your computer to turn it into an MP3? Well, Grasshopper, try this little USB audio interface. A phono cartridge needs a pre-amp with special equalization. The USB Phono Plus has true phono inputs — along with line-level and digital audio connections — in a solid metal enclosure. It works with standard USB audio drivers on any recent Win or Mac system, and includes a wall wart so you can use it as a standalone phono pre-amp.

My favorite feature? It has a *knob*. The gain control lets you fine-tune the volume before digitizing, to record the level as high as possible without clipping. But trust your recording software's level meters; my clip LED on the front panel flashed red even when there was a bit of headroom to spare. No complaints about the sound quality, though. I can't hear any difference between my original vinyl and the digitized version.

—Ross Orr

Photography by Howard Cao

3

3 5-Gallon Bucket Organizer

Bucket Boss 56 $20
duluthtrading.com

▪ Who doesn't love a white 5-gallon bucket? I've personally seen 'em used as beer coolers, impromptu stools, ashtrays, boombox holder-uppers, paint mixers, clam chowder storinators, and — during one troubling night in a forest primeval — I saw one installed with a garbage bag and used as an ersatz camp toilet. Poor little bucket.

The good folks at Bucket Boss built an entire company around things you can do to mod your plain old 5-gallon bucket. Their flagship product is the Bucket Boss 56 — a super-dense, water-resistant fabric sleeve that slips over the lip of the bucket to provide a geeky embarrassment of pockets for all the tools and gadgets in your world. That's 56 pockets to be precise — 38 running around the outside, plus 18 more inside. It even has a duct tape dispenser and a big-ass holster for your drill. Rawr.

Knowing they've got a good thing on their hands here, the Bucket Bosses offer a variety of additional products, including a compartmentalized lid for storing screws and nuts, a wheeled dolly for pushing the bucket around, a cushioned grip for the handle — even a model with a little cushion that turns your bucket into a comfy-looking seat. They also sell something they call the Mug Hugger, a Bucket Boss scaled down to fit on a mug for organizing your pens, pencils, and scissors, and, "in the cause of modesty and good taste," the Longtail T, designed to make plumber's butt a thing of the past.

While some of this aftermarket frippery may turn off your stauncher bucket puritans, MAKE enthusiasts can appreciate the extreme range of happy hacks gladly accepted by our humble 5-gallon friend. If containers were software, the bucket would be GNU/Linux, and the Bucket Boss 56 would be its cleverest value-added reseller.

—*Merlin Mann*

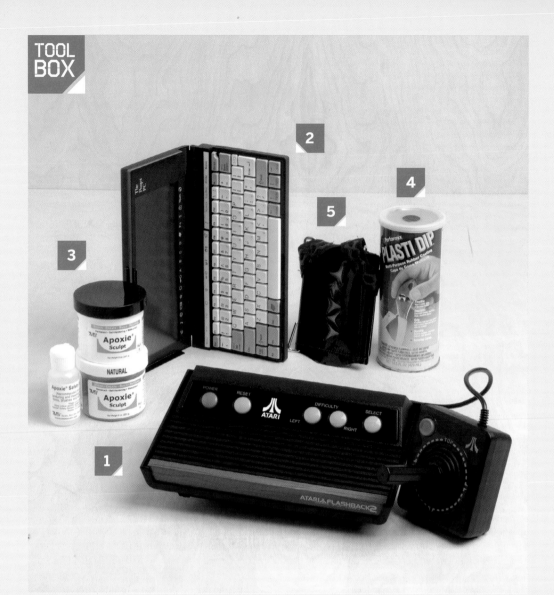

A Better (and Hackable) Retro Atari

Atari Flashback 2 $30
atari.com

- The first Atari Flashback retro game unit was a disappointment. Worst of all, its processor was similar to (blasphemy!) the Nintendo Entertainment System's. Thank the gaming gods, the Atari Flashback 2 gets it right. Not only does it look like a smaller version of the 2600, its internals recreate the 2600 processor, and its classic joysticks actually work better than the originals (using modern-day circuitry).

The Flashback 2 comes with 40 built-in games, including old-time favorites *Adventure, Missile Command*, and *Yar's Revenge*, a few Activision titles, and brand-new games made specifically for it.

The Flashback 2 was even purposely designed to tempt one to hack it. Though it doesn't come with a cartridge slot, plans detailing how to add one have been released on the web along with some other hacks by an engineer who developed the hardware (atarimuseum. com/fb2hacks). The fact that you can mod the Flashback 2 to reuse your old Atari 2600 cartridges is its coolest feature of all.

—Howard Wen

2 An Inexpensive New/Old Word Processor

Poqet PC $30
cadigital.com/poqetpc.htm

▪ Some things in the computer industry were so far ahead of their time that they did not gather a following until years after the parent company quit doing business. First sold in 1991, the Poqet "pocket" computer measured just 9"x5"x1" when closed and sported near-state-of-the-art technology for its time. The Poqet PC ran DOS 3.3 (on ROM), had a 25-line, 80-character display and a keyboard large enough for touch typing, and could run for up to 8 hours on two AA cells. It stored programs and files on a pair of 1MB RAM disks, and accessories included a 3.5" floppy drive and an RS232 serial port.

Originally retailing for over $2,000, it still has a following among folks looking for a low-cost, lightweight, basic word processing tool. Refurbished units can be found on the internet, marketed by California Digital, Inc.   —*Thomas J. Arey*

3 Epoxy for Big Projects

Apoxy Sculpt $11
avesstudio.com

▪ Most people are familiar with plumber's epoxy putty — mix two parts together, cures rock hard in 20 minutes. Unfortunately, it's usually sold in small quantities at a relatively high price and has a short working time. Recently, I was thrilled to discover a world of epoxy putties I never knew existed; you can buy large quantities at a reasonable price and in more flavors than a Baskin-Robbins. Apparently, this is the same stuff used by professional modelers for Disney displays, museums, and movie props.

I originally bought it for modeling, but I am constantly amazed at how useful it is for other tasks, from using it as insulating material for a hot wire foam cutter to replacing large chunks of painted wood trim eaten by our dog. The epoxy putties from Aves Studio are nontoxic, stick to almost anything, won't shrink, and dry rock hard. They can be sanded, drilled, tapped, frozen, heated … you name it.

Now, I can finally finish my Flying Spaghetti Monster idol and do some serious worshiping.

—*Mark Lengowski*

4 Multipurpose Rubber Coating

Plasti Dip $8
plastidip.com

▪ Plasti Dip International has a range of maker-friendly rubber coating products (like Super Grip Fabric Non-Skid Fabric Coating and Liquid Tape Electrical Brush-On Insulation), but my favorite is the classic Plasti Dip, a gooey substance in a can.

To rubber-coat an object, you lower it into the can and then remove it and allow it to dry (you can also paint the rubber coating onto a surface). The flexible rubber coating works great to add a better gripping surface to a tool, or to insulate an object from electrical shock or vibration. To create a great non-skid coating, you can mix three parts Plasti Dip to one part pumice grit.

Plasti Dip is easy to use, and easy to clean up if you goof; let it dry, and you can just peel it off. (The bit I spilled on the floor cleaned right up.)

—*Terrie Miller*

5 Bond Girl

Stylish Leg Purse $79
tsaya.com

▪ Doesn't everybody want a secret pocket on their person to store useful gadgets? Having just watched a sleep-inducing cigarette procured from a garter case in *The Spy Who Loved Me*, I have been hankering after a similar accessory. To my delight, the Tsaya, a sleek, patent-leather purse that straps inconspicuously around your thigh, seems to fit the bill.

My first foray is at a dinner party, where I have impetuously fastened it over jeans. Aside from having a poor grip on denim, the strap-over-pants look seems to alarm dinner guests. Worn under a black skirt, it's more successful. At a San Francisco nightspot, fellow clubbers are intrigued; one girl asks if you can use it like an iPod armband, and someone else wants to know if it comes in lace.

Rock on! While its appearance could be enhanced for a broader audience (its thick straps and shiny patent leather are more bondage gear and less Coach accessory), the appeal of carrying cellphones and other vitals hands-free more than makes up for it. Now I just have to find those 007 cigarettes — if anyone has a source, let me know.

—*Jenna Phillips*

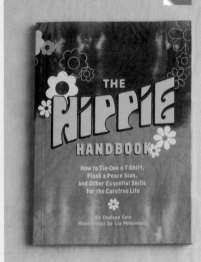

PALLET BIKE RAC

Ah, the humble pal
our great consume
getting slimed and
These old salts ha
don't deserve to be
side—especially w
bike racks! Set on
enough to hold a v
to do a little sand
rims). Just prop it
and paint for a me
permanent place

1 Ready to Make

***ReadyMade How to Make {Almost} Everything: A Do-It-Yourself Primer*, by Shoshana Berger and Grace Hawthorne** $25, Clarkson Potter

▪ This whimsical instructional guide from the folks at *ReadyMade* magazine (a sort of sister publication to MAKE) shows you how to make useful objects from things formerly destined for the dustbin: a ceiling fan constructed from an old bike wheel and a plastic bowl, a wine rack rigged from coat hangers, a dog bed fashioned from an old pair of blue jeans, a rug assembled from shopping bags, a picture frame made out of a hardback book. Half the joy of this book came from seeing how the authors breathed new life into unwanted junk. The other half came from the way it inspired me to start thinking of novel uses for the neglected stuff in my own garage.

—*Mark Frauenfelder*

2 The Hippie Handbook

How to Tie-Dye a T-Shirt, Flash a Peace Sign, and Other Essential Skills for the Carefree Life, by Chelsea Cain; illustrations by Lia Miternique $12.95, Chronicle Books

▪ Offering up *The Hippie Handbook* to your friends quickly becomes a sociological study. The younger hippies — say, 35 and under who still live more or less by the hippie constructs — will most likely love this paperback (with a tie-dye cover, natch).

But the older hippies — or even worse, those fortunate (?) few who were raised by hippies — seem to dislike the book, or are at least likely to rip its tips to shreds. To this group, being a hippie is no laughing matter; a childhood full of homemade yogurt and granola can clearly make a person angry later in life.

Regardless of who your friends are, the book is a blast. You'll read about everything from how to make "oregano" brownies to how to care for a fern, how to milk a goat, and the proper way to do tree sitting (bring a laptop). The illustrations are charming, and really add to the Flower Child feeling of the book. Peace and love. —*Shawn Connally*

Bill's Excellent Adventure

Adventures from the Technology Underground,

by **William Gurstelle** $25, Crown Publishing

■ To find out what happens at Teslathon, LDRS (Large Dangerous Rocket Ships), Burning Man, or the World Championship Punkin Chunkin, you wouldn't want to rely on mainstream media. A typical reporter struggles not only to understand these technophiles but to explain how things work, trusting that few people actually care. Not Bill Gurstelle.

In his new book, *Adventures from the Technology Underground*, Gurstelle takes us inside the communities of amateur inventors and science enthusiasts.

He spent two years researching what he calls the "Technology Underground," a counter-culture based on "self-directed science and technology." Gurstelle told me that "the media usually portrayed these people as either beer-drinking sods with a welding torch or uber-geek technology freaks. Of course, mostly they were just people who enjoyed technology."

Like Brian Basura, who built a very large electrostatic discharge machine, designed to shrink quarters in his garage. Gurstelle writes that high-voltage tinkerers also know "how to make a big-time, lethally dangerous high-voltage capacitor in an unfinished basement; and they can make it from a 12-pack of empty Rolling Rock beer bottles wrapped in aluminum foil and immersed in brine."

The author of *Backyard Ballistics*, a mix of history and how-to, Gurstelle not only introduces us to fascinating people and the places they gather, but offers straightforward explanations for the technologies that obsess them, such as high-power amateur rocketry, pulsejets, tesla coils, catapults, and flamethrowers. He helps us see these technophiles as they see themselves. —*Dale Dougherty*

How Your Birding Hobby Can Help Further Science

eBird Online Database Free
ebird.org

For centuries, backyard scientists have kept lists of the birds they've seen. These lists have been useful to scientists who are able to correlate the observed changes in range or migration times of species with global climate change. Patterns from shorter-term effects, like West Nile virus, can also emerge.

The Cornell Lab of Ornithology and the Audubon Society joined forces to create an easy-to-use online database, eBird, that lets you track your bird observations and pool them with others, so the data can be used by anyone.

With eBird, your personal birding list becomes more interesting and fun to browse right away. But by sharing your data, you also become a citizen scientist. The citizen scientists of the 1800s didn't realize that their observations would be important clues about global warming, and the data you collect now may be important in ways we can't even foresee. —*Terrie Miller*

Collaborative Music Playing on the Web

NINJAM Free
ninjam.com

Not only does NINJAM enable musicians to play together online, it even saves the jams as Creative Commons-licensed recordings. But the breakthrough is the way it overcomes the groove-destroying delay inherent in sending audio over the net; it delays each performer's output by an interval of several bars. Strange concept, but it actually encourages listening.

I downloaded the free NINJAM client (available for Windows, Mac, and Linux), logged in, and started playing electric piano. Suddenly, a guitarist chimed in, followed by vocals. NINJAM co-founder Justin Frankel (of Winamp, Gnutella, and Shoutcast fame) even joined on drums.

I experienced a few audio glitches and dropouts, but the mix seems to have made it to the website intact. This is one rough ninja, but it's free and fun. —*David Battino*

An iTunes Alternative for the Shuffle

Sean Shrum's Shuffler Free
shuffle-db.sourceforge.net

The iPod Shuffle is probably the biggest-selling Flash-based MP3 player in the world. Given that you can drag and drop files onto it from any decent file manager (on Windows, Mac, Linux), you would think you could just copy your MP3 files onto the Shuffle and listen to them.

Unfortunately, Apple wants you to keep using iTunes. Which is where Sean Shrum's Shuffler application comes in handy. It's a small application that will scan all the files on your Shuffle and rebuild the internal music database.

You still need to use iTunes to set up your Shuffle as a USB Memory Drive, although this is a one-time operation. After that, all you have to do is just copy your music over from any host machine, run the database builder, and away you go. —*Ewan Spence*

Rechargeable Batteries That Really Work

1

Rayovac Rechargeable Battery
Charger $20 rayovac.com

▪ My whiz-bang digital camera consumes batteries at an alarming rate. Conventional wisdom is to use rechargeable AA cells, but they seemed to always be dead when I wanted to use the camera. So, I either put in disposables or waited a couple of hours while they recharged. Who knows what the Pulitzer committee has missed as a result?

The problem is the relatively high self-discharge rate of nickel metal hydride (NiMH) batteries, which is a problem for gadgets like my camera that go weeks between uses. I found a good workaround

in Rayovac's NiMH batteries, which are available in AA or AAA sizes. These batteries each have a tiny internal charging circuit and sensors that allow them to be fully charged in a blistering 15 minutes. Their 2000mAh capacity matches standard NiMHs, and they are otherwise interchangeable with their not-so-smart brothers.

Now I can recharge my camera in the time it takes to find my shoes and pack my bag. The Pulitzer guys still haven't called, though.

—*Bob Scott*

2   A Really Good Universal Remote

Logitech Harmony 880 Universal Remote $249
logitech.com

▪ The joy of the Harmony 880 is in the setup wizard, which is part of a personal online account you create on Logitech's website. In the wizard, you select devices you want to control (like your DVD player) by manufacturer and product number, and then define activities (watch a DVD), which could include your receiver, video projector, and DVD player.

Then you install the remote software on to your computer, connect the Harmony 880 through a USB cable, and download the setup to your remote. You will want to set up the activities, even if all you want to do is turn two devices on and flip through channels. If this sounds like a lot, it really isn't. As you first use the remote, there is an onscreen assistant to fix any problems that appear. This is one seriously smart remote.   *—Dane Picard*

3   Maximum Hard Drive Portability

CoolMax CD-311 Hard Drive Chassis $60
coolmaxusa.com

▪ I have a fascination with versatility — the more ports and switches a device has, the better. When looking for a hard drive chassis, then, the CoolMax CD-311 was an easy choice. As an on-site consultant, I'm often called upon to do emergency data transfers and system recoveries. The CD-311 will support whatever drive I'm working with, whether it comes out of an eMac or an Xserve.

The unit's external interfaces are also myriad, supporting FireWire, USB 2.0, and SATA (with cables included for all three). Internal cables for both IDE and SATA are included as well, not to mention a small Phillips screwdriver.

The design lends itself well to being cut in half, which would allay my fears about heat dissipation in the fanless, sealed aluminum chassis. It would be very easy to make a portable drive dock similar to WiebeTech's ComboDock — but while the ComboDock with SATA adapter costs $250, the CD-311 is under $60. Hacksaw not included.   *—Tom Owad*

4   Sticky Situation

Self-Fusing Rubber Splicing Tape $11
digikey.com, part #3M130C-ND

▪ If you do any kind of electrical work or home repairs, you likely rely on heat-shrink tubing and/or electrical tape to seal and insulate wiring splices. But even the best vinyl electrical tape will eventually disintegrate into sticky black goo, and heat-shrink tubing needs a heat source. (Plus, if you are like me, you never have the right diameter handy.)

The pros have a solution: like other sticky miracles, it comes from 3M, and it's called "Linerless Rubber Splicing Tape." It fuses to itself to form a nearly perfect waterproof bond. To use it, you cut a short piece off the roll, stretch to activate it, and tightly wrap your splice. The result is a neatly insulated tube of tough black rubber. It also works great on anything that needs protective wrapping or a better grip, such as elderly extension cords or sports racquet handles. I even use it to bind rope ends so they don't unravel.   *—Jonathan Foote*

5   Superior Nonstick Pans

8-Piece Calphalon One Cookware Set $479
calphalon.com

▪ Warranties are essential for expensive purchases, and the warranty on my set of Calphalon One cookware is phenomenal. Unless you submerge the set into an acid bath or melt it in lava, you're pretty much covered for life.

My favorite feature of the cookware is the incredible nonstick surface. While all humans may be created equal, all nonstick surfaces are not. I now cook eggs, pancakes, vegetables, and other items using a fraction of the oil I did with my previous so-called "nonstick" set. I'm also a fan of the "stay cool" dual handles on the large pieces. These are especially useful for the chef's skillet and the large pot, which get pretty heavy when you fill them up.

It's also really nice to be able to put this stuff in the oven — even the broiler — and see it come out unscathed. Ever melted a handle or two while preheating the oven? I won't be having that problem anymore.   *—Matthew Russell*

The Alpha Maker

Four Seasons of *MacGyver* on DVD

$113 amazon.com

▪ **Creating an impromptu smoke-screen, plugging an engine hole with egg whites, and dismantling a missile with a paperclip were all in a typical day's work for television's fictional alpha maker, MacGyver. With the recent release of the complete first four seasons on DVD, we now get to relive all of these amazing and innovative MacGyverisms.**

As a kid, the show inspired me to invest several weeks unsuccessfully trying to create explosives out of pine cones, but I feel it was time well spent. MacGyver is one of the few shows that doesn't make me feel dumber after watching it, and that's the main reason I love it so much.

—Matthew Russell

A Table Saw for Hackers

Ryobi BT3100 10" Table Saw $299

ryobitools.com

▪ The Ryobi BT3100 10" table saw is accurate, inexpensive, and cleverly engineered. Its accuracy is due to a number of well-designed features, although the design may make some traditionalists a bit uncomfortable. There are no miter slots (though most people find the sliding miter table more than sufficient for all tasks), and the rip fence and sliding miter table are aligned to the blade.

The rip fence is rock-solid; it locks down and stays locked down. This allows for such a feat as ripping ¹⁄₆₄-inch strips from a 6-foot work piece. Blade run-out is negligible due to an efficient, belt-driven arbor design, which allows the saw to cut a standard 4x4 in one pass.

The BT3100 is a unique saw and sure to keep any-one with a tight budget and a small workspace very satisfied. It also has a loyal following — big enough for not one, but two, online forums.

For more information, see bt3central.com and ryobitools.com.   —*Mark Lengowski*

T.J. "SKIP" Arey N2EI is a freelance writer and author of *Radio Monitoring: The How-To Guide.*

Dave Battino plays Mac, PC, and keyboards. More at batmosphere.com.

Jonathan Foote is a research scientist in Silicon Valley.

Mark Lengowski spends way too much time making things instead of earning an honest living.

Merlin Mann helps people make interesting things for the Global Interweb.

Ross Orr keeps the analog alive in Ann Arbor, Mich.

Tom Owad is the author of *Apple I Replica Creation* and editor of applefritter.com.

Jenna Phillips is currently launching Formula Magic, a high-tech shoe company.

Dane Picard is a video art-ist and filmmaker based in Los Angeles.

Matthew Russell tries hard to live life as a renaissance man, but is distracted by the cult of Mac.

Bob Scott is a statistical construct of various con-sumer electronics market-ing departments.

Ewan Spence keeps it real in Scotland. His blog is ewanspence.com.

Howard Wen (howardwen. com) is a frequent contribu-tor to MAKE.

Have you used something worth keeping in your toolbox? Let us know at toolbox@makezine.com.

Prewar Maker's Primer
By H.B. Siegel

Amateur Craftsman's Cyclopedia of Things to Make

■ **It would be hard to overestimate the** number of hours I pored over this book as a kid. Published in 1937 by Popular Science Publishing Company, Inc., it was obsolete decades before I was born, yet it is captivating in the glimpse it provides into the pre-WWII maker's world. At 330-plus pages and one to five projects per page, it is filled with projects that range from simple to elaborate, from relevant to bizarre. A few are even politically insensitive in our era.

I did not come from a "maker" household, so I was restricted to window-shopping. I was fascinated by the hand-drawn illustrations and black-and-white photographs. No Illustrator or Photoshop for another 50 years, thank you! The gorgeous "Rakish Privateer Clipper of 1812" was intricate beyond all hope, requiring milling, woodwork, and fittings. But the "Mayan Throwing Sticks" looked possible, and the kids in the illustration were having a great time.

While several of the projects make little sense today, such as "Comical Coconut Bird Serves as an Ash-tray" or "Eight Scoop Popcorn Server," there are hundreds that are still appropriate and approachable, e.g., "Lathe Gear Cutting," "Glass Cutting Jig Removes Bottle Tops," "Homemade Furnace Melts Aluminum," "Homemade Tool Shapes Metal Tubing," and "An Easy Way to Turn True Wooden Balls."

It was the games and toys that drew me in again and again — the "Walking Top Game," the "Magic Knives for Trick Surgery," the "Rubber Bulb Toy Submarine," and especially the "Ticket-Dropping Airplane," which offered assurances to "start a party off with a bang." I had no lack of toys as a kid, but it was mind-blowing that someone could *make* these.

The scary ones are fun as well. No dust masks, gloves, safety goggles, saw guards, or warnings in sight. Were some lives shortened while "Making a Mercury Barometer," or a "Simple Hygrometer Made with Carbon Tetrachloride," or the double-whammy of "Smoker's Set Made of Thin Sheet Lead"? What are we blithely making now that will horrify our grandchildren?

There's something for everyone in this book. It demonstrates recycling and reuse way before it was cool. Recommended.

⬆ **Instructions for making a small furnace that can melt aluminum, good for making castings of "attractive novelties."**

⬅ Copies of *Amateur Craftsman's Cyclopedia of Things to Make* appear frequently on Amazon and eBay for about $10.

➡ **Make sure to recite an "appropriate airplane verse" before setting this tiny payload-delivering aircraft aloft over the heads of a "gathering of young folks ... The resulting confusion and laughter will break the ice and start the party off in the right spirit of hilarity."**

H.B. Siegel is technology director for IMDb.com (part of Amazon.com). He has been CTO for ILM and in management at Pixar, Wavefront, and SGI.

Images courtesy of H.B. Siegel

Computers in the Machine
By Tom Owad

Emulation software gives you free, zero-footprint, vintage computers.

■ **I've had a spate of flakey hardware** recently. A wire-wrap board that only works when it's flexed, an SE/30 that randomly refuses to boot, and an Apple IIgs that only occasionally recognizes its hard drive have all prompted me to take a closer look at emulation. Space, too, has played a role. While classic computers overflow the garage, the emulators merely occupy a directory on my hard drive. The manuals and disks for them, instead of filling shelves, reside in PDFs and disk images.

Thanks to Mini vMac, my Mac Plus rarely comes off the shelf. I no longer worry about wear and tear on 20-year-old floppy disks. They've been imaged to my PowerBook and safely locked away. The emulator makes it possible to quickly inspect old programs for my website, without the painstaking task of physically transferring them from a Power-Book G4 to a Mac Plus.

All of my classic Mac games are now just a few clicks away, but as anybody knows, playing in an emulator just isn't the same as having the real thing. Granted, it's much more convenient when there's serious work to be done, but for just playing around, the hardware is half the fun. There are some systems, though, where the hardware is either too expensive or too large to make ownership practical. The DEC PDP-8, a classic mini-computer from the 60s and 70s, is a good example. Simply finding a PDP-8 is a significant undertaking, let alone amassing the necessary peripherals (paper tape reader, RK05 disks, etc.) and getting them all running.

PDP-8/E Simulator, by Bernhard Baehr, does a fantastic job of bringing the PDP-8 to life on your desktop, offering more functionality than the original. While the 8/E has 4K words of memory by default, the simulator comes with a plugin for a KM8-E Memory Extension, bringing the memory up to 32K words (each word is 12 bits). There are additional plugins for an ASR 33 Teletype, PC8-E High-Speed Paper Tape Reader, RK8-E Disk Cartridge System, LP8-E Line Printer, three different clocks, and a TSC8-75 board (for running the ETOS time-sharing operating system). There's even a KC8-EA Programmer's Console (i.e., "front panel") so you can toggle switches and watch the blinkenlights.

The API for plugins is well documented and included with the source code. With it you can write emulators for additional I/O devices or even create new devices of your own. The simulator is also well suited for development. I hear PDP-8 development isn't quite the burgeoning industry it used to be, but its educational value remains. Each device (plugin) has a window displaying its internal state. Through the PDP-8/E CPU window, it is possible to view and edit each register and the contents of memory. The CPU supports breakpoints, a trace mode, and many other features you'd expect to find in an advanced simulator. The front panel plugin is fully functional, so if you prefer the classic switches-and-lights view, it is entirely possible to edit memory, load programs, and single-step the processor using it.

> "PDP-8/E Simulator does a fantastic job of bringing the PDP-8 to life on your desktop, offering more functionality than the original."

In fact, Baehr's website contains a tutorial explaining how to operate a PDP-8 with no system software, by using the front panel. The process involves toggling in an 18-line program for reading paper tape (a paper tape reader and punch is built into the ASR 33 Teletype you have attached). This program is then used to load a more sophisticated paper tape reader, which you can finally use to load your own program.

For more sophisticated use, attach an RK8-E Disk System, and you can run Focal-8, OS/8, Pascal-S, and ETOS. Mount the RK05 DECpack that contains your desired system, toggle in the RK8-E boot code, and run. OS/8 was the most common operating system for the PDP-8 and is a good place to start. Included on the OS/8 DECpack is the classic text editor TECO, from which Emacs eventually developed. TECO documents are created by the OS/8 command "MAKE filename." For a glimpse at 1970s hacker humor, try entering the command "MAKE LOVE."

Tom Owad (owad@applefritter.com) is a Macintosh consultant in York, Pa., and editor of Applefritter (applefritter.com). He is the author of *Apple I Replica Creation* (Syngress, 2005).

Saturday, April 22nd and Sunday, April 23rd

Maker Faire

Meet the Makers

■ Join the MAKE team and thousands

of other makers at MAKE magazine's first ever Maker Faire — inspiration, know-how, networking, and just plain fun for makers and aspiring makers of all ages and backgrounds. It's the first ever public gathering of tech DIY enthusiasts, crafters, educators, tinkerers, hobbyists, science clubs, students, authors, and exhibitors.

It's two days with six exposition and workshop pavilions, a 5-acre outdoor midway, over 100 exhibiting makers, hands-on workshops, demonstrations, DIY competitions, and the latest in tools, kits, and DIY resources.

It's Weird Science, Ultimate Garage, Robotics, Digital Entertainment/Gaming, Green Tech & Electronics Recycling, Ultimate Workshop, and the MAKE: Remix video film festival.

Bring your family, friends, and a student or three to the San Mateo Fairgrounds (centrally located in the San Francisco Bay Area) for a weekend of hands-on exploration, recipe-sharing, creative mischief-making, and wholesome play. All-day admission is a paltry $12 for adults, $5 for teens, and not a cent for kids 12 and under accompanied by an adult. Maker-friendly family tickets and full weekend passes are available as well!

Information and registration for the following can be found at makezine.com/faire:

Tickets: 50% off weekend passes for MAKE readers when purchased by March 31. Register and enter "MFWINTER" as your access code.

MAKE: Remix Video Film Festival: Contest details available online.

To Exhibit as a Maker: Further details and proposal information for makers interested in exhibiting opportunities are available online.

To Sponsor or Exhibit at Maker Faire: Please contact Larry Ecklund, the director of event sales at lecklund @makezine.com.

■ The Silly Subscription Renewal Game

We're mad as hell, and we're not going to take it anymore! The magazine industry bombards us with renewal reminders, but by the time our subscription actually runs out, we get an incredible offer and a gift tossed in to boot. They've trained us to hold out, cluttering our mailboxes and wasting paper and money.

So here's our promise to you: Your best renewal offer from MAKE will always come up front. If you play the hold-out game, chances are you'll pay more. Sorry, but we think subscribers who cost us less ought to pay less.

We've also put together an auto-renew program. When it's time to renew, we'll send an email letting you know that we're renewing your subscription and charging your card. You'll always receive at least $5 off the standard subscription price, you'll receive just one email, and you'll have access to the entire MAKE archive when you access the MAKE Digital Edition, available only to MAKE subscribers.

To register for auto-renewal, please visit makezine.com/premier.

■ The Gift of MAKE

Buy a gift subscription for someone (yes, even if you are that special someone) and get a free MAKE T-shirt for yourself. Visit makezine.com/gift.

➡ MAKE: The First Year Boxed Set Collector's Edition
For the most dedicated maker who wants to keep the complete first year of MAKE in pristine condition, we are offering a special 4-volume gift-box collector's edition. Just need the cool collector's box? We have that too (makezine.com/go/coolstuff).

⬅ Makers: The Book
Steal away, kick your feet up, and lose yourself in the creativity and resourcefulness of the maker movement with this beautiful hardbound book. Bob Parks and the editors of MAKE profile 100 makers and their homebrew projects, backyard inventions, and basement creations (also at makezine.com/go/coolstuff).

Our favorite events from around the world — March - April - May 2006

March

>> **Space Shuttle STS: 121 Launch**, March 4, Cape Canaveral, Fla. The space shuttle *Discovery* is scheduled to launch into orbit at 3:21 p.m. EST. This mission will test new equipment and procedures to increase shuttle safety and will deliver more supplies and cargo for future space station expansion. kennedyspacecenter. com/launches/schedule Status.asp

>> **EDITOR'S CHOICE — Nevada Test Site: Public Access Day**, March 23, April 26, and May 24, Las Vegas. The U.S. Department of Energy provides tours of the Nevada Test Site only a few times a year. The 250-mile bus tour begins in Las Vegas and takes visitors to nuclear test locations such as Frenchman's Flat, Mercury, and Control Point 1. nv.doe.gov/nts/tours.htm

>> **The Texas Mile: Land Speed Racing Event**, March 19–20, Goliad, Texas. Land speed racers from around the country meet on the long, long, flat runway of the Goliad Industrial Air Park to test their fastest motorcycles and automobiles. Some machines attain speeds of more than 250 mph. texasmile.com

April

>> **Edinburgh Science Festival**, April 6–16, Edinburgh, Scotland. A large and diverse series of science-oriented lectures, tours, and demonstrations, this world-class festival brings together many of the top thinkers and great science entertainers of the day in the U.K.'s largest public celebration of science and technology. sciencefestival.co.uk

>> **Hacker Con Notacon**, April 7–9, Cleveland. This annual conference explores and showcases the technologies, philosophy, and creativity associated with hacker life. notacon.org

>> **Trinity Site: Public Access Day**, April 8, White Sands Missile Range, N.M. Trinity Site was the first "ground zero": the spot where the first atomic bomb was placed on a 100-foot steel tower and detonated on July 16, 1945. Open to the public just two days a year. www.wsmr.army.mil/pao/ TrinitySite/trndir.htm

>> **13th Annual Trinity College Fire Fighting Home Robot Contest**, April 8–9, Hartford, Conn. This event features a robot fire fighting contest, a Robotics Olympiad, and a Robotics Symposium. trincoll.edu/events/robot

>> **12th Annual Mobot Races**, April 21, Carnegie Mellon University, Pittsburgh. Student teams compete for prize money as they race autonomous, self-powered miniature robots on a torturous, curving, 255-foot-long course. cs.cmu.edu/~mobot

>> **Thunder Over Louisville**, April 22, Louisville, Ky. Each spring, 500,000 people gather to watch what is said to be the country's largest regularly scheduled pyrotechnics display. TOL also includes an air show with over 100 aircraft. It takes place just before the Kentucky Derby. thunderover louisville.org/show/ default.asp

>> **FIRST 2006 Robotics Championship**, April 27–29, Atlanta. The annual FIRST Robotics Championship is the final step in a multinational robotics competition that involves 25,000 people, nearly 1,000 teams, and 30 preliminary competitions. Open to the public, free of charge. usfirst.org/robotics

May

>> **EDITOR'S CHOICE — NRO L-26 Delta 4-Heavy Launch**, May 1, Cape Canaveral, Fla. The biggest lifter in the U.S. rocket arsenal, the Boeing Delta 4-Heavy rocket, launches a classified spy satellite cargo on May 1. This rocket features three booster cores mounted together to form a triple-body rocket. More power and flames than anywhere else on the planet. kennedyspacecenter.com/ launches/scheduleStatus. asp

>> **Astronomy Day**, May 6. Astronomy Day events take place across the United States. Stargazing, lectures, and demonstrations. astroleague.org/al/ astroday/astroday.html

>> **The Intel International Science and Engineering Fair**, May 7–13, Indianapolis. The world's largest pre-college science competition, with over $3 million in awards and scholarships, will attract 1,500 top science high school students from the 50 U.S. states and 40 countries. sciserv.org/isef

>> **Andrews Air Force Base Joint Services Open House**, May 20–21, Washington, D.C. This event pulls in crowds of nearly 400,000 people, mainly for the spectacular aerial demonstrations, including precision aerial maneuvers by the Air Force's Thunderbirds, fly-bys, parachute jumping, and ground displays. public.andrews. amc.af.mil/index.asp

>> **Jet Propulsion Laboratory Open House**, May 20–21, Pasadena, Calif. This event showcases JPL's accomplishments with exhibits and demonstrations about the laboratory's ongoing research and space exploration. Many of the lab's scientists and engineers will be on hand to answer questions. www.jpl.nasa. gov/pso/oh.cfm

>> **EDITOR'S CHOICE — World Championship 38th Annual Kinetic Sculpture Race**, May 27–29, Arcata to Ferndale, Calif. Art and engineering combine to produce imaginative human-powered racing machines. Thousands watch as vehicles resembling an 85-foot salmon, a giant chocolate éclair, and a yellow submarine race for the finish line during the long and grueling race. kineticsculpturerace.org/

IMPORTANT: All times, dates, locations and events are unconfirmed and subject to change. Verify all information before making plans to attend.

Do you know of an event that should be included? Please use our online form at makezine.com/events. Sorry, it is not possible to include listings for all events in the magazine, but they will be listed online.

Did you attend an interesting event? If so, please tell us about it in our Forums (forums.makezine.com).

Where makers tell their tales and offer praise, brickbats, and swell ideas.

The MAKE blog is the only site I check 3-5 times a day and actually enjoy.

MAKE made me realize that I'm not just someone with the hacker mindset — I'm someone who wants to create things myself. I never had an interest in crocheting, but the fact that I want to make a tux scarf makes me realize that things are just better when you make them yourself.   —Troy Fletcher

I saw MAKE this morning. I subscribed immediately after seeing the VCR cat feeder in Vol. 3. I proceeded to check out the site, and haven't done a lick of web work since.

In fact, I will probably eat a large amount of chemical enhancements, shave my head, and devote my existence to building things from MAKE and/or coming up with at least one project to submit.

A haiku to express my newfound obsession with MAKE:

Client site is down,
I am owned by the makezine.
Where's my flamethrower?

—Benjamin Jones

I've been subscribing to MAKE since Vol. 2. As you can imagine, my subscription renewal is coming due. I was considering letting my subscription lapse, but Vol. 4 changed my mind. I think it was your first Really Good issue. Why? Because it finally had articles and projects I'm interested in.

I liked the projects on music bending, high-speed photography, turning a cup into a speaker, programming microcontrollers (though I don't understand everyone's fascination with the PIC; in my opinion the AVR chip is just as inexpensive and has a much more orthogonal [i.e. more consistent] instruction set), and the article about Dean Kamen. My personal interest is in electronic projects and robotics. Taking apart existing products and repurposing them is cool.

Your recent online list of $100-and-under gift ideas for a maker was spot-on. Loved it.

Can we see more articles about embedded processors and Linux devices?

—Barry Brown

 Mister Jalopy,

You can't imagine how happy (and a bit disappointed) I was to find your site ... especially the LP ripping Apple/iPod 1960s radio. As you can see [from the photo], I thought I was pretty smarty pants when I built a networked MP3 Touchscreen Jukebox out of a 1940 RCA chassis that I bought for Can$40 ... but yours blows mine completely out of the water. ;-)

I wanted to say how much I enjoy your site and attitude toward the garage life and taking pleasure in cleaning, restoring, and recycling items that most folks would throw out! I read your article on welding in MAKE magazine, and I have to say it was well done. I am actually finishing my Arc welding class as we speak, so I am really looking forward to buying my first welder and getting set up ... I have been more of a woodworker up to now!

Anyways. I am now a convert to your site and want you to keep up the great work ... if ever you want a guest gadgeteer from the Great White North, let me know. I would be very happy to oblige.   —Patrice Collin

I owe you a debt of gratitude. Your magazine saved Christmas. I was a Grinch beyond anything Theodore Geisel could have imagined. I hated Christmas; I was tired of the rampant consumerism and sterilized joy that comes with only giving DVDs and video games to friends and family. But thanks to you, Christmas is fun again. After finally getting my subscription to MAKE and then getting all the back issues. I could only think of one thing to give on Christmas — marshmallow guns! I tinkered with the design a little, and gave it an amusing name, the Mallow Mauser. I handed out the guns disassembled; ten in all, with instructions, and in the confusion of the moment sprang a surprise attack on the family. The guns were put together fast, and thanks to a controlled leak on my part, safety glasses were already on hand. My family is made up of mechanics, engineers, and contractors and they all loved it. The adults definitely had more fun than their kids. To explain where this idea came from, I brought my collection of MAKE magazines. Two out of three uncles now say they are getting a subscription to MAKE. —*Mike Ciarlone*

■ I just want you to know that I am deeply, completely, and hopelessly in love with your magazine. I didn't think I could be any more in love until I logged on to your digital version. Now I realize that was only puppy love. This is the real thing. Please never stop.
—*Suzanne Portnoy*
Mother, Wife, Yoga Student,
PTA Member, Closet Hacker

■ When I was about 11 years old, my grandparents gave me a gift subscription to *National Geographic World*. Though I'd already been pretty heavily into the grownup *NG*, I was delighted with that kids' version, not least because it included how-to instructions that got me making stuff. The one I remember most vividly was an issue that had me building, understanding, and using a pinhole camera (aluminum foil on sticky paper for the aperture). I wish I still had those images, though the memories of the experience, shared with my photographer father, are more valuable by far.

When each new issue of MAKE arrives, I react, if possible, with more excitement than an 11-year-old receiving a long-awaited gift. Probably 75% of the stories in any given issue cover projects I would never actually presume to undertake, but that's not the point. Just knowing that there are people out there building aerial-photography kites, light-seeking robotic mice, or VCR-powered cat feeders, hacking this, that, and everything in between — this knowledge, the sheer range of possibility that your magazine deals with, and the sense that I could do it all, too — fills me with joy, and I devour every issue with the enthusiasm of an 11-year-old waking up to the world.

I also appreciate the editorial stance on digital rights. Your information about how to tweak my European DVD player so it will play what I own, where I own it, has been worth the price of admission alone. I've renewed my subscription. I hope to see you grow and prosper in Year 2.
—*J.D. Stephens*

■ I love MAKE & war not? Anyway, I love the online edition and plan to subscribe as a Christmas gift to myself and get (not make) my wife Fran to order the back issues for me. It is very comforting to know there are other subversives out there like myself, although I can't hold a candle to most of the skills displayed in your wonderful articles.
—*Kevin P. Wilkinson*

■ I love your blog on makezine.com and you guys are doing a great job with the magazine. I was glad to get the digital copy of the last issue since mail service here in New Orleans leaves something to be desired since Katrina hit. —*Clay McGovern*

MAKE AMENDS

In Volume 04, page 21, through some inexplicable turn of events, an old and inaccurate version of the "Unroll Your Own" story ran. The 2004 "Stuck at Prom" winners featured in the photograph are Kris Murray and Caitlyn Waters, from Oklahoma City, Okla.

In Volume 04, page 47, a line of text was left out. The sentence should have read: "It's way more fun than the Porsche, and it carries far more failed experiments to the beach," he said. The best thing about meeting the weirdoes is that the conversation works to communicate information first, NDAs later, if ever."

In Volume 04, page 130, the schematic for wiring the anonymous megaphone is drawn incorrectly. The diode and trimpot connections are reversed, and the polarities were left off the capacitors. A revised schematic can be downloaded at makezine.com/go/revised.

In Volume 04, pages 158-165, the Primer incorrectly uses the term "BASIC Stamp" to generically describe the products of Parallax and netMedia. The article incorrectly tells the reader that the "BASIC Stamp" is any full-blown module with a micro, memory, power supply, etc., and also incorrectly suggests that PICs are available from Microchip, Atmel, etc. In reality, PICs are only available from Microchip, and AVRs are available from Atmel.

Pushing Electrons
By Phillip Torrone

Bots, how-tos, and virtual worlds all on the MAKE: Blog. Take a tour of what's new with Associate Editor Phillip Torrone.

◀ MAKE, Volume 03, displayed in a virtual library inside Second Life.

■ We have been pushing a lot of

atoms and electrons since we kicked off MAKE, and before we knew it, the MAKE: Blog hit 3,000 posts in less than a year — during the first week of 2006. That's about ten posts a day, all about the amazing things makers are doing, building, and creating. And we're just getting warmed up: 24/7, makers are sending us their stories, their works, and their ideas. You, too, can visit Makezine.com and just clickity-click the submit button (makezine.com/blog).

We've also added a few new features and ways to get what you need from MAKE and to publish your projects.

First up is the MAKEbot. The MAKEbot is an instant message buddy you add to your AOL/AIM/iChat buddy list. He can tell you the latest news, photos, bookmarks, and forum posts, as well as search all of MAKE from your buddy list or from your phone. We're constantly making the MAKEbot smarter, so give it a try. Just type "help" to get a list of all the commands!

Next, we're pleased to announce that we've partnered up with Instructables. What the heck is an Instructable? Instructables is a step-by-step collaboration system that helps you record and

share your projects with a mixture of images, text, CAD files, ingredient lists, and more. Head on over and add your projects to the MAKE group (instructables.com/make).

Then there's our MAKE Flickr photo pool. Have a hack, mod, tweak, or just something plain cool you want to share? Join in, upload, and share your photos with hundreds of makers around the world (flickr.com/groups/make/pool).

Forums! You asked for 'em, we got 'em! We launched the forums on Makezine.com, so anyone can talk about all things MAKE-related — from getting help in projects to selling MAKE wares in the Maker Marketplace. If you've ever posted on MAKE in the comments, your account works for the forums too, or just sign up and start posting with other makers.

Lastly, visit MAKE in the virtual world. We've created the first magazine based on a real magazine inside the virtual world of Second Life (secondlife.com). It's free to download and use for PCs and Macs, and a Linux version is coming soon at secondlife://Tenera/64/192.

Phillip Torrone is associate editor of MAKE.

Image courtesy of Torrone Trumbo

Create Explosive Visuals
By Chris Kenworthy

Real movie explosions are dangerous, expensive, and usually unnecessary. You can create a convincing explosion with well-lit water, and add it to your footage later. You need to choose a good sound effect to make this work well; check the internet.

You will need: A large, clean mirror, camera, towels, black balloon.

1 The balloon used in this example is red, but you should use a black balloon. A black balloon won't show up in the finished shot, ensuring that only the brightly lit water is visible. To create the illusion of a powerful explosion, fill the balloon halfway with water, and then add enough air to fill it up until it is close to bursting. This combination of air and water yields a good explosion, but you can experiment with other ratios.

2 To create this effect, you need to work in a dark, enclosed space. Lean the mirror against a wall or over a box, at a 45-degree angle to the floor. Suspend the balloon over the mirror. Your movie light should be positioned off to the side, aimed at the balloon. Keep the light far enough away that it won't be splashed when the balloon explodes.

When you stand in front of the mirror, you should be able to see the balloon. (It shows up clearly in this example because the balloon is red.) You'll find it easier to line up your camera on the balloon if you tape a brightly colored piece of paper to it. Remove it when you're ready to shoot.

3 Zoom in on the mirror so that nothing is visible except the balloon. Use the zoom rather than moving the camera forward to avoid any risk of getting water on the camera.

4 Now pop the balloon. Use black tape to attach a sharp nail or tack to a black stick, which won't be visible when you move it into frame. Start shooting and pop your balloon. When it pops, brightly lit water will explode toward the camera.

5 In your editing software, layer the explosion footage over the object you want to blow up. Use a Color Correction filter to shift the colors toward yellow. If the object is moving, click the Keyframe button and drag the explosion layer so that it matches the movement of the object. Keep explosion shots brief, and cut them in with crowd reactions or people running away.

Excerpted from *Digital Video Production Cookbook*, ISBN 0596100310 (O'Reilly Media, Inc.).

Photography courtesy of Chris Kenworthy

GEORGE DYSON

A Treehouse Grows in British Columbia

Three years, 95 feet above the Earth.

■ **David Brower taught me the laws** of mountaineering. First: "Climbing is safer than staying home." Second: "Never step on anything you can step over, and never step over anything you can step around." When I left the mountains of California to become a boat builder in British Columbia, I recast this wisdom, in honor of one of the legendary scavengers of the Vancouver waterfront, as Jim Land's Law: "Never buy anything you can make, and never make anything you can find."

In 1972, at age 19 and facing my third winter in Canada, I built a small treehouse 95 feet up in a Douglas fir, and lived there for three years. This adventure began by accident, when the boat I was

> "Never buy anything you can make, and never make anything you can find."

deckhanding on hit a large cedar log in Georgia Strait. We towed the log back with us to our anchorage within Vancouver's inner harbor, at the mouth of Indian Arm. As I began to split the log into shakes, I started to think of building a small cabin for the winter — like Malcolm Lowry, who had written his masterpiece *Under the Volcano* while squatting in a shack on the other side of the inlet, now Cates Park.

In the 1970s, the line between legitimate and illegitimate tenancy remained indistinct. Many coastal residents lived in houses built on floats, and when someone slid a house onto the shoreline (almost entirely "Crown land") no one paid much notice — or any tax. The Queen of England (also the Queen of Canada) rarely came by to check. One more cabin in the woods would probably not attract attention, but the rainforest was dark and damp. How about going up?

There happened to be one large, landmark Douglas fir right at the water's edge. The first set of branches were 30 feet up, and that would have been the logical place to build a house. But I kept climbing, and the view just got better and better, until, from near the top of the tree I had a panoramic view from Second Narrows to the 6,565-foot summit of Mount Meslillooet, directly north. There was a clear path through the branches from there to the ground, so I installed a pulley, found a 200-foot length of nylon

Photograph by Ann E. Yow

◀ **George Dyson's tree-house, constructed 95 feet above the ground in the branches of a Douglas fir on the shores of Burrard Inlet at Belcarra Park, British Columbia, in 1972.**
➡ **Oliver Thomas, George Dyson, and Claudia Thomas in the doorway of the treehouse. Too short to climb from branch to branch, the children ascended by rope.**

rope, and started hauling things up.

For the initial framing, I used alder poles, lashed with two pounds of #15 tarred nylon net-mending twine. This and a length of stovepipe were the only materials I bought to build the house. Fourteen living branches were incorporated into the structure, with all the panels triangulated so that the house could flex in the wind but not twist itself apart.

Once the structure was framed, I installed a floor of salvaged boards, and then began covering the roof with cedar shakes and shingling the outer walls (this requiring some precarious, self-belayed rope work). The house had four windows — two made from tempered glass salvaged from junked console televisions and two octagonal windows salvaged from a remodeled house — and a "Trout" wood-burning stove. Along the other side, supported by one of the internal branches, I built a bunk with firewood storage underneath.

The November gales whistled through the shake walls and roof, so I insulated them with slabs of styrofoam packaging, and then paneled the entire house with split cedar boards on the inside. The cedar grain was so fine (almost 100 growth rings per inch) that some of the boards above my bunk spanned 700 years. During the winters I spent in the treehouse, I read the journals of Cook, Bering, La Perouse, Galiano, Vancouver, and the other early visitors to the Northwest Coast, and pondered how the entire recorded history of British Columbia and Southeast Alaska had only penetrated the outer few inches of the log I had collided with.

I climbed the tree in all conditions, day or night, and never had any misadventure going up, and only one misadventure coming down. I usually rappelled, using the doubled firewood-hauling line, one end clipped into a climber's descending harness and the other end through an aluminum brake. I would leave the house in near free-fall, and then after about 60 feet, start applying the brake. One day, my long, scraggly beard caught in the brake on the way down. *"Merde!"* as they say in Quebec.

I never imagined the treehouse would survive as well as it did. It withstood torrential rains, snow, ice, and winds gusting to 60 knots. During storms, the house gyrated eight or ten feet. When people think of the rainforest, they think damp moss, luxuriant fungi, and rotting wood. But that's on the ground; in the forest canopy, the rain is just passing by. Even though I would leave the house for weeks at a time, I never returned to find dampness, mildew, or mold. The sun hit my windows long before those of anyone else living on the slopes of Indian Arm, and I enjoyed a microclimate completely different from that on the ground.

The three winters I spent in the treehouse were the most memorable of my life. I had no political agenda, nor was I a recluse: I needed a place to live, I liked climbing, and I even entertained visitors now and then. More people should build more treehouses, and we need more Crown land. Sure, at 95 feet, it may have been a little dangerous, but you can get killed walking across the street. As Brower would say, it was probably safer than staying on the ground.

George Dyson, a kayak designer and historian of technology, lives in Bellingham, Wash., and is the author of *Baidarka, Project Orion,* and *Darwin Among the Machines.*

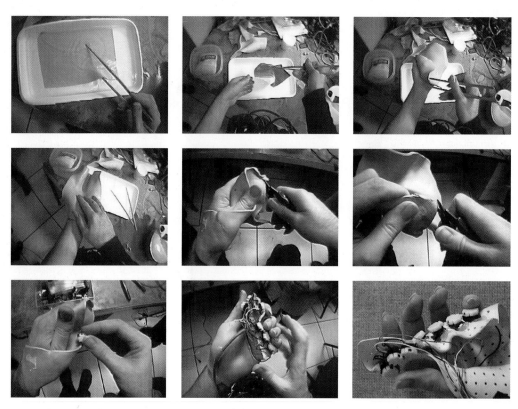

HOMEBREW
My Multiambic Keyer
By Steve Mann

Back in the 1970s, I wanted to be able to type while walking around, or compose music while jogging, so I came up with various devices that were like a keyboard but without the board.

Of these, one of my favorites was a cluster of keys mounted to a piece of plastic that I'd molded to exactly fit my hand.

One of the simplest designs is the pentambic keyer, having 5 switches, one for each finger and thumb, resulting in 17,685 different combinations.

The septambic keyer provides a slight improvement by using 3 thumb switches, resulting in more than 29 million possible combinations, making it possible to map one English word to each possible combination.

In addition to merely typing text, I also use the keyer to compose music, so that I can generate notes in time to my footsteps. This results in the ability to produce music at a nice constant tempo, in time with a steady walk or jog. (I tend to keep jogging in place when a stoplight turns red.)

Periodic structures, like the granite slabs in a public square, or stairs, result in an enhanced sense of timing. For example, the stairs in my building have 9 steps per flight (half floor), so if I always take 3 steps per landing and I get 6 bars of music (in 4/4 time) per floor, going up 3 floors I can play exactly the first verse of a Gershwin lullaby ("Summertime, and the livin' is easy ...").

If I miss a beat, I know right away because I'll end up with an "off-by-one" error when I get to the top of the stairs. Instead of merely using switches, I use pressure sensors, so that the harder I squeeze, the louder the note or chord that is sounded, and thus, the keyer gives rise to a very expressive musical instrument.

If you would like to build a keyer for typing text, or a "musikeyer" for composing music while you jog, you can learn more from wearcam.org/septambi/index.html and from my textbook, *Intelligent Image Processing*, published by John Wiley and Sons.

Email Steve Mann at mann@eecg.toronto.edu.

Photography courtesy of Steve Mann